Learn

Eureka Math™

Grade 3
Modules 3 & 4

Published by Great Minds®.

Copyright © 2018 Great Minds®

Printed in the U.S.A.

This book may be purchased from the publisher at eureka-math.org.

10 9 8 7 6 5 4 3 2 1

v1.0 PAH

ISBN 978-1-64054-061-3

G3-M3-M4-L-05.2018

Learn ◆ Practice ◆ Succeed

Eureka Math™ student materials for *A Story of Units*® (K–5) are available in the *Learn, Practice, Succeed* trio. This series supports differentiation and remediation while keeping student materials organized and accessible. Educators will find that the *Learn, Practice,* and *Succeed* series also offers coherent—and therefore, more effective—resources for Response to Intervention (RTI), extra practice, and summer learning.

Learn

Eureka Math Learn serves as a student's in-class companion where they show their thinking, share what they know, and watch their knowledge build every day. *Learn* assembles the daily classwork—Application Problems, Exit Tickets, Problem Sets, templates—in an easily stored and navigated volume.

Practice

Each *Eureka Math* lesson begins with a series of energetic, joyous fluency activities, including those found in *Eureka Math Practice*. Students who are fluent in their math facts can master more material more deeply. With *Practice*, students build competence in newly acquired skills and reinforce previous learning in preparation for the next lesson.

Together, *Learn* and *Practice* provide all the print materials students will use for their core math instruction.

Succeed

Eureka Math Succeed enables students to work individually toward mastery. These additional problem sets align lesson by lesson with classroom instruction, making them ideal for use as homework or extra practice. Each problem set is accompanied by a Homework Helper, a set of worked examples that illustrate how to solve similar problems.

Teachers and tutors can use *Succeed* books from prior grade levels as curriculum-consistent tools for filling gaps in foundational knowledge. Students will thrive and progress more quickly as familiar models facilitate connections to their current grade-level content.

Students, families, and educators:

Thank you for being part of the *Eureka Math*™ community, where we celebrate the joy, wonder, and thrill of mathematics.

In the *Eureka Math* classroom, new learning is activated through rich experiences and dialogue. The *Learn* book puts in each student's hands the prompts and problem sequences they need to express and consolidate their learning in class.

What is in the Learn *book?*

Application Problems: Problem solving in a real-world context is a daily part of *Eureka Math*. Students build confidence and perseverance as they apply their knowledge in new and varied situations. The curriculum encourages students to use the RDW process—Read the problem, Draw to make sense of the problem, and Write an equation and a solution. Teachers facilitate as students share their work and explain their solution strategies to one another.

Problem Sets: A carefully sequenced Problem Set provides an in-class opportunity for independent work, with multiple entry points for differentiation. Teachers can use the Preparation and Customization process to select "Must Do" problems for each student. Some students will complete more problems than others; what is important is that all students have a 10-minute period to immediately exercise what they've learned, with light support from their teacher.

Students bring the Problem Set with them to the culminating point of each lesson: the Student Debrief. Here, students reflect with their peers and their teacher, articulating and consolidating what they wondered, noticed, and learned that day.

Exit Tickets: Students show their teacher what they know through their work on the daily Exit Ticket. This check for understanding provides the teacher with valuable real-time evidence of the efficacy of that day's instruction, giving critical insight into where to focus next.

Templates: From time to time, the Application Problem, Problem Set, or other classroom activity requires that students have their own copy of a picture, reusable model, or data set. Each of these templates is provided with the first lesson that requires it.

Where can I learn more about Eureka Math *resources?*

The Great Minds® team is committed to supporting students, families, and educators with an ever-growing library of resources, available at eureka-math.org. The website also offers inspiring stories of success in the *Eureka Math* community. Share your insights and accomplishments with fellow users by becoming a *Eureka Math* Champion.

Best wishes for a year filled with aha moments!

Jill Diniz

Jill Diniz
Director of Mathematics
Great Minds

The Read–Draw–Write Process

The *Eureka Math* curriculum supports students as they problem-solve by using a simple, repeatable process introduced by the teacher. The Read–Draw–Write (RDW) process calls for students to

1. Read the problem.

2. Draw and label.

3. Write an equation.

4. Write a word sentence (statement).

Educators are encouraged to scaffold the process by interjecting questions such as

- What do you see?

- Can you draw something?

- What conclusions can you make from your drawing?

The more students participate in reasoning through problems with this systematic, open approach, the more they internalize the thought process and apply it instinctively for years to come.

Contents

Module 3: Multiplication and Division with Units of 0, 1, 6–9, and Multiples of 10

Module 4: Multiplication and Area

Grade 3
Module 3

Geri brings 3 water jugs to her soccer game to share with teammates. Each jug contains 6 liters of water. How many liters of water does Geri bring?

_____

_____

_____

_____

Read   Draw   Write

Name _____   Date _____

1. a. Solve. Shade in the multiplication facts that you already know. Then, shade in the facts for sixes, sevens, eights, and nines that you can solve using the commutative property.

×	1	2	3	4	5	6	7	8	9	10
1		2	3							
2		4		8				16		
3						18				
4					20					
5										50
6		12								
7										
8										
9										
10										

b. Complete the chart. Each bag contains 7 apples.

Number of Bags	2		4	5	
Total Number of Apples		21			42

2. Use the array to write two different multiplication sentences.

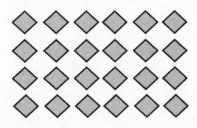

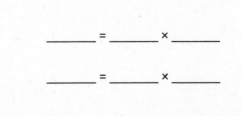

_____ = _____ × _____

_____ = _____ × _____

Lesson 1:   Study commutativity to find known facts of 6, 7, 8, and 9.

5

3. Complete the equations.

a.   2 sevens = _____ twos

  = ___14___

b.   3 _____ = 6 threes

  = _____

c.   10 eights = 8 _____

  = _____

d.   4 × _____ = 6 × 4

  = _____

e.   8 × 5 = _____ × 8

  = _____

f.   _____ × 7 = 7 × _____

  = ___28___

g.   3 × 9 = 10 threes − _____ three

  = _____

h.   10 fours − 1 four = _____ × 4

  = _____

i.   8 × 4 = 5 fours + _____ fours

  = _____

j.   _____ fives + 1 five = 6 × 5

  = _____

k.   5 threes + 2 threes = _____ × _____

  = _____

l.   _____ twos + _____ twos = 10 twos

  = _____

Lesson 1:   Study commutativity to find known facts of 6, 7, 8, and 9.

EUREKA
MATH™

Name _____   Date _____

1. Use the array to write two different multiplication facts.

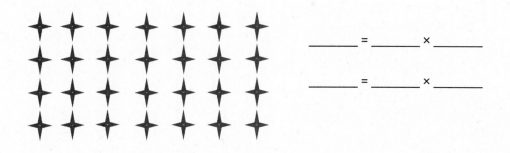

_____ = _____ × _____

_____ = _____ × _____

2. Karen says, "If I know 3 × 8 = 24, then I know the answer to 8 × 3." Explain why this is true.

Jocelyn says 7 fives has the same answer as 3 sevens + 2 sevens. Is she correct? Explain why or why not.

_____

_____

_____

_____

Read   Draw   Write

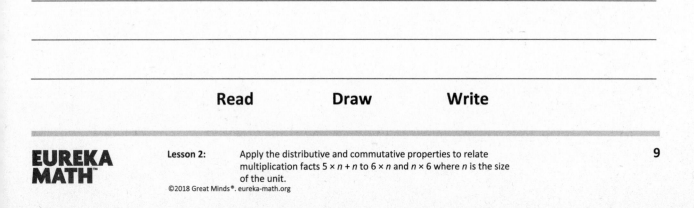

Name _____   Date _____

1. Each 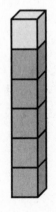 has a value of 7.

Unit form: 5 _____

Facts: 5 × _____ = _____ × 5

Total = _____

Unit form: 6 sevens = _____ sevens + _____ seven

= 35 + _____

= _____

Facts: _____ × _____ = _____

_____ × _____ = _____

Lesson 2:   Apply the distributive and commutative properties to relate
multiplication facts 5 × n + n to 6 × n and n × 6 where n is the size
of the unit.

©2018 Great Minds®. eureka-math.org

11

2. a. Each dot has a value of 8

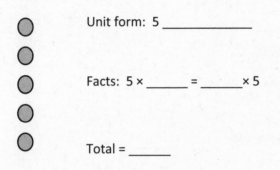

Unit form: 5 _____

Facts: 5 × _____ = _____ × 5

Total = _____

b. Use the fact above to find 8 × 6. Show your work using pictures, numbers, or words.

Lesson 2:   Apply the distributive and commutative properties to relate
                 multiplication facts 5 × n + n to 6 × n and n × 6 where n is the size
                 of the unit.

**EUREKA
MATH**™

3. An author writes 9 pages of her book each week. How many pages does she write in 7 weeks?
   Use a fives fact to solve.

4. Mrs. Gonzalez buys a total of 32 crayons for her classroom. Each pack contains 8 crayons. How many packs of crayons does Mrs. Gonzalez buy?

5. Hannah has $500. She buys a camera for $435 and 4 other items for $9 each. Now Hannah wants to buy speakers for $50. Does she have enough money to buy the speakers? Explain.

EUREKA
MATH™

Lesson 2:   Apply the distributive and commutative properties to relate multiplication facts 5 × n + n to 6 × n and n × 6 where n is the size of the unit.

©2018 Great Minds®. eureka-math.org

13

Name _____   Date _____

Use a fives fact to help you solve 7 × 6. Show your work using pictures, numbers, or words.

Lesson 2:   Apply the distributive and commutative properties to relate
              multiplication facts 5 × *n* + *n* to 6 × *n* and *n* × 6 where *n* is the size
              of the unit.

©2018 Great Minds®. eureka-math.org

15

Twenty-four people line up to use the canoes at the park. Three people are assigned to each canoe. How many canoes are used?

_____

_____

_____

_____

Read   **Draw**   **Write**

Lesson 3:   Multiply and divide with familiar facts using a letter to represent the unknown.

Name _____   Date _____

1. Each equation contains a letter representing the unknown. Find the value of the unknowns, and then write the letters that match the answers to solve the riddle.

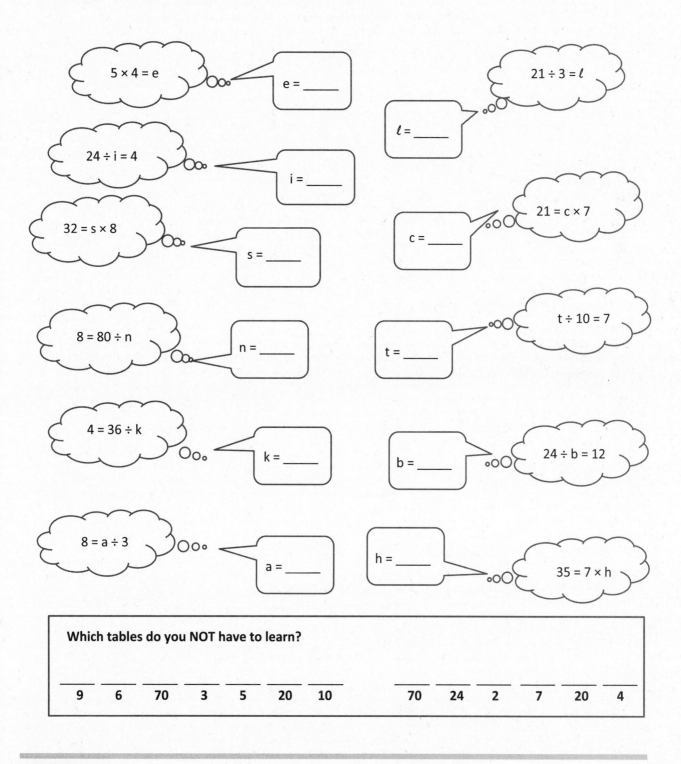

$5 \times 4 = e$ $e =$ _____

$21 \div 3 = l$ $l =$ _____

$24 \div i = 4$ $i =$ _____

$21 = c \times 7$ $c =$ _____

$32 = s \times 8$ $s =$ _____

$t \div 10 = 7$ $t =$ _____

$8 = 80 \div n$ $n =$ _____

$4 = 36 \div k$ $k =$ _____

$24 \div b = 12$ $b =$ _____

$8 = a \div 3$ $a =$ _____

$35 = 7 \times h$ $h =$ _____

Which tables do you NOT have to learn?

___ ___ ___ ___ ___ ___ ___
9   6   70   3   5   20   10

___ ___ ___ ___ ___ ___
70   24   2   7   20   4

2. Lonna buys 3 t-shirts for $8 each.

   a. What is the total amount Lonna spends on 3 t-shirts? Use the letter *m* to represent the total amount of money Lonna spends, and then solve the problem.

   b. If Lonna hands the cashier 3 ten dollar bills, how much change will she receive? Use the letter *c* in an equation to represent the change, and then find the value of *c*.

Lesson 3: Multiply and divide with familiar facts using a letter to represent the unknown.

3. Miss Potts used a total of 28 cups of flour to bake some bread. She used 4 cups of flour for each loaf of bread. How many loaves of bread did she bake? Represent the problem using multiplication and division sentences and a letter for the unknown. Then, solve the problem.

_____ × _____ = _____

_____ ÷ _____ = _____

4. At a table tennis tournament, two games went on for a total of 32 minutes. One game took 12 minutes longer than the other. How long did it take to complete each game? Use letters to represent the unknowns. Solve the problem.

CHALLENGE!

Lesson 3:   Multiply and divide with familiar facts using a letter to represent the unknown.

21

©2018 Great Minds®. eureka-math.org

Name _____   Date _____

Find the value of the unknown in Problems 1–4.

1. $z = 5 \times 9$

    z = _____

2. $30 \div 6 = v$

    v = _____

3. $8 \times w = 24$

    w = _____

4. $y \div 4 = 7$

    y = _____

5. Mr. Strand waters his rose bushes for a total of 15 minutes. He waters each rose bush for 3 minutes. How many rose bushes does Mr. Strand water? Represent the problem using multiplication and division sentences and a letter for the unknown. Then, solve the problem.

    _____ × _____ = _____

    _____ ÷ _____ = _____

Lesson 3:   Multiply and divide with familiar facts using a letter to represent
the unknown.

©2018 Great Minds®. eureka-math.org

23

Marshall puts 6 pictures on each of the 6 pages in his photo album. How many pictures does he put in the photo album in all?

_____

_____

_____

_____

Read   **Draw**   **Write**

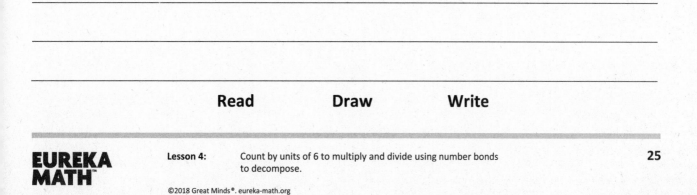

Lesson 4:   Count by units of 6 to multiply and divide using number bonds
              to decompose.

©2018 Great Minds®. eureka-math.org

25

Name _____   Date _____

1. Skip-count by six to fill in the blanks. Match each number in the count-by with its multiplication fact.

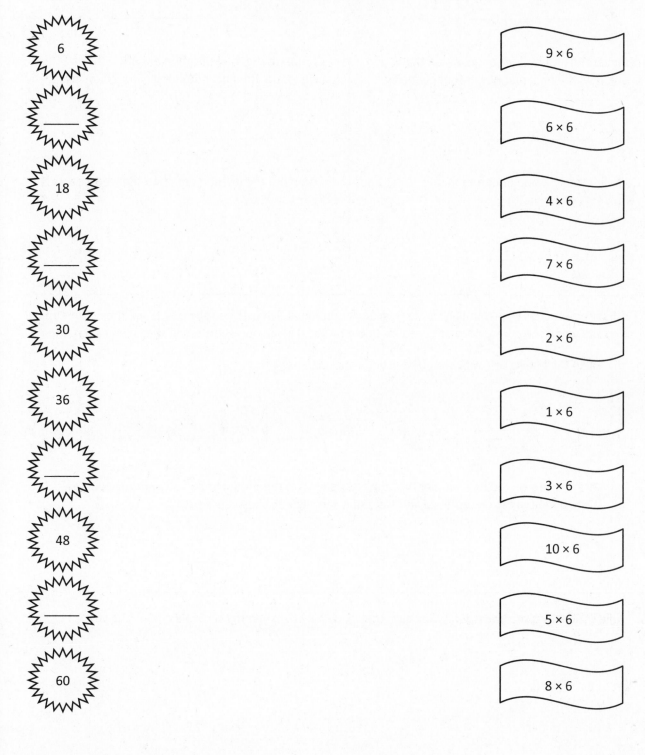

EUREKA MATH

Lesson 4:   Count by units of 6 to multiply and divide using number bonds to decompose.

27

©2018 Great Minds®. eureka-math.org

2. Count by six to fill in the blanks below.

6, _____, _____, _____

Complete the multiplication equation that represents the final number in your count-by.

6 × _____ = _____

Complete the division equation that represents your count-by.

_____ ÷ 6 = _____

3. Count by six to fill in the blanks below.

6, _____, _____, _____, _____, _____, _____

Complete the multiplication equation that represents the final number in your count-by.

6 × _____ = _____

Complete the division equation that represents your count-by.

_____ ÷ 6 = _____

4. Mrs. Byrne's class skip-counts by six for a group counting activity. When she points up, they count up by six, and when she points down, they count down by six. The arrows show when she changes direction.

   a. Fill in the blanks below to show the group counting answers.

   ↑ 0, 6, _____, 18, _____ ↓ _____, 12 ↑ _____, 24, 30, _____ ↓ 30, 24, _____ ↑ 24, _____, 36, _____, 48

   b. Mrs. Byrne says the last number that the class counts is the product of 6 and another number. Write a multiplication sentence and a division sentence to show she's right.

   6 × _____ = 48   48 ÷ 6 = _____

5. Julie counts by six to solve 6 × 7. She says the answer is 36. Is she right? Explain your answer.

Lesson 4:   Count by units of 6 to multiply and divide using number bonds to decompose.

EUREKA MATH™

Name _____   Date _____

1. Sylvia solves 6 × 9 by adding 48 + 6. Show how Sylvia breaks apart and bonds her numbers to complete the ten. Then, solve.

2. Skip-count by six to solve the following:

   a. 8 × 6 = _____   b. 54 ÷ 6 = _____

Gracie draws 7 rows of stars. In each row, she draws 4 stars. How many stars does Gracie draw in all? Use a letter to represent the unknown and solve.

Read   **Draw**   **Write**

Lesson 5: Count by units of 7 to multiply and divide using number bonds to decompose.

31

Name _____   Date _____

1. Skip-count by seven to fill in the blanks in the fish bowls. Match each count-by to its multiplication expression. Then, use the multiplication expression to write the related division fact directly to the right.

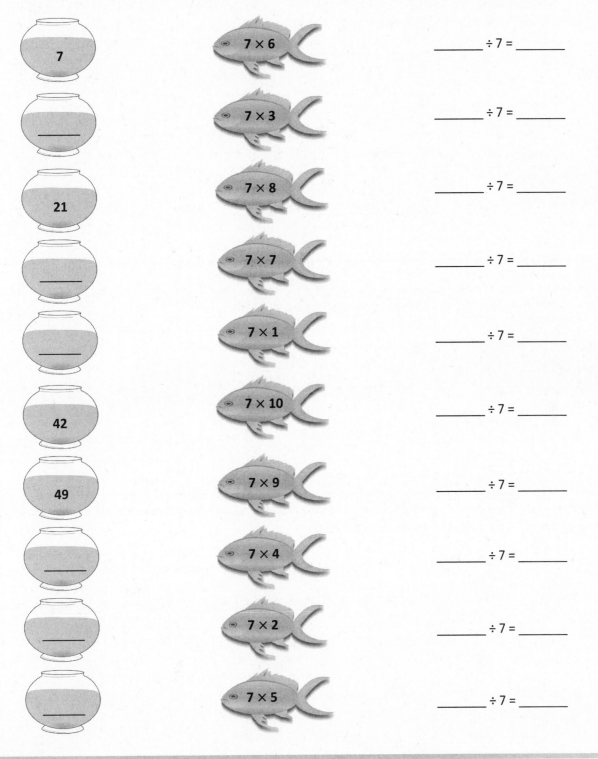

_____ ÷ 7 = _____

_____ ÷ 7 = _____

_____ ÷ 7 = _____

_____ ÷ 7 = _____

_____ ÷ 7 = _____

_____ ÷ 7 = _____

_____ ÷ 7 = _____

_____ ÷ 7 = _____

_____ ÷ 7 = _____

_____ ÷ 7 = _____

2. Complete the count-by seven sequence below. Then, write a multiplication equation and a division equation to represent each blank you filled in.

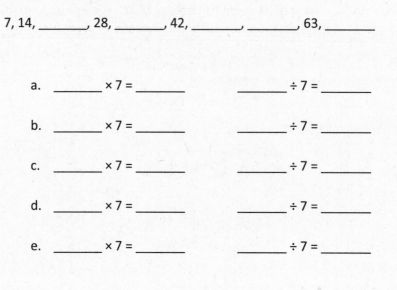

7, 14, _____, 28, _____, 42, _____, _____, 63, _____

    a.   _____ × 7 = _____   _____ ÷ 7 = _____

    b.   _____ × 7 = _____   _____ ÷ 7 = _____

    c.   _____ × 7 = _____   _____ ÷ 7 = _____

    d.   _____ × 7 = _____   _____ ÷ 7 = _____

    e.   _____ × 7 = _____   _____ ÷ 7 = _____

3. Abe says 3 × 7 = 21 because 1 seven is 7, 2 sevens are 14, and 3 sevens are 14 + 6 + 1, which equals 21. Why did Abe add 6 and 1 to 14 when he is counting by seven?

4. Molly says she can count by seven 6 times to solve 7 × 6. James says he can count by six 7 times to solve this problem. Who is right? Explain your answer.

Lesson 5:   Count by units of 7 to multiply and divide using number bonds to decompose.

©2018 Great Minds®. eureka-math.org

Name _____   Date _____

Complete the count-by seven sequence below. Then, write a multiplication equation and a division equation to represent each number in the sequence.

7, 14, _____, 28, _____, 42, _____, _____, 63, _____

a. _____ × 7 = _____   _____ ÷ 7 = _____

b. _____ × 7 = _____   _____ ÷ 7 = _____

c. _____ × 7 = _____   _____ ÷ 7 = _____

d. _____ × 7 = _____   _____ ÷ 7 = _____

e. _____ × 7 = _____   _____ ÷ 7 = _____

f. _____ × 7 = _____   _____ ÷ 7 = _____

g. _____ × 7 = _____   _____ ÷ 7 = _____

h. _____ × 7 = _____   _____ ÷ 7 = _____

i. _____ × 7 = _____   _____ ÷ 7 = _____

j. _____ × 7 = _____   _____ ÷ 7 = _____

Lesson 5:   Count by units of 7 to multiply and divide using number bonds
             to decompose.

©2018 Great Minds®. eureka-math.org

35

Mabel cuts 9 pieces of ribbon for an art project. Each piece of ribbon is 7 centimeters long. What is the total length of the pieces of ribbon that Mabel cuts?

_____

_____

_____

_____

Read   **Draw**   **Write**

EUREKA
MATH™

Lesson 6:   Use the distributive property as a strategy to multiply and divide
              using units of 6 and 7.

©2018 Great Minds®. eureka-math.org

37

Name _____   Date _____

1. Label the tape diagrams. Then, fill in the blanks below to make the statements true.

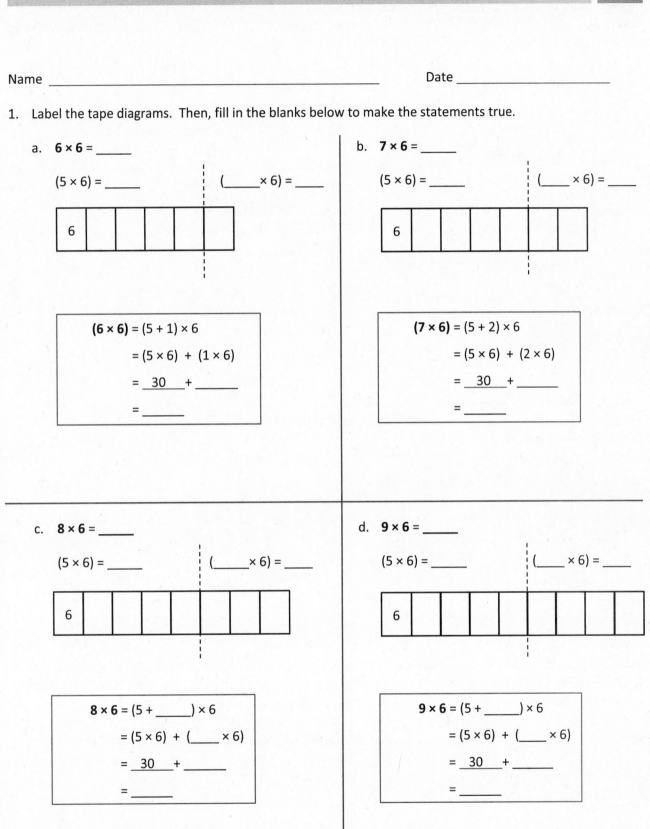

a. **6 × 6** = _____

(5 × 6) = _____   (_____ × 6) = _____

| 6 | | | | | |

(6 × 6) = (5 + 1) × 6

= (5 × 6) + (1 × 6)

= __30__ + _____

= _____

b. **7 × 6** = _____

(5 × 6) = _____   (_____ × 6) = _____

| 6 | | | | | | |

(7 × 6) = (5 + 2) × 6

= (5 × 6) + (2 × 6)

= __30__ + _____

= _____

c. **8 × 6** = _____

(5 × 6) = _____   (_____ × 6) = _____

| 6 | | | | | | | |

8 × 6 = (5 + _____) × 6

= (5 × 6) + (_____ × 6)

= __30__ + _____

= _____

d. **9 × 6** = _____

(5 × 6) = _____   (_____ × 6) = _____

| 6 | | | | | | | | |

9 × 6 = (5 + _____) × 6

= (5 × 6) + (_____ × 6)

= __30__ + _____

= _____

Lesson 6:   Use the distributive property as a strategy to multiply and divide   39
             using units of 6 and 7.

©2018 Great Minds®. eureka-math.org

2. Break apart 54 to solve 54 ÷ 6.

3. Break apart 49 to solve 49 ÷ 7.

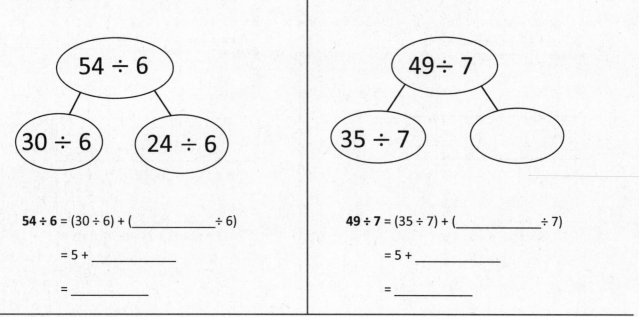

54 ÷ 6 = (30 ÷ 6) + (_____ ÷ 6)

= 5 + _____

= _____

49 ÷ 7 = (35 ÷ 7) + (_____ ÷ 7)

= 5 + _____

= _____

4. Robert says that he can solve 6 × 8 by thinking of it as (5 × 8) + 8. Is he right? Draw a picture to help explain your answer.

5. Kelly solves 42 ÷ 7 by using a number bond to break apart 42 into two parts. Show what her work might look like below.

Lesson 6:   Use the distributive property as a strategy to multiply and divide using units of 6 and 7.

©2018 Great Minds®. eureka-math.org

Name _____   Date _____

1. A parking lot has space for 48 cars. Six cars can park in 1 row. Break apart 48 to find how many rows there are in the parking lot.

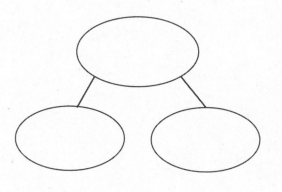

_____

2. Malia solves 6 × 7 using (5 × 7) + 7. Leonidas solves 6 × 7 using (6 × 5) + (6 × 2). Who is correct? Draw a picture to help explain your answer.

Lesson 6:   Use the distributive property as a strategy to multiply and divide
using units of 6 and 7.

41

©2018 Great Minds®. eureka-math.org

Name _____ Date _____

1. Match the words to the correct equation.

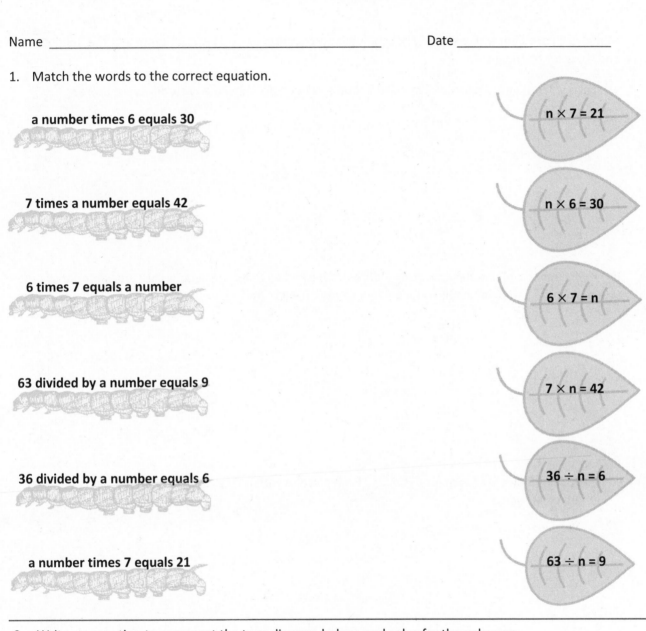

a number times 6 equals 30

7 times a number equals 42

6 times 7 equals a number

63 divided by a number equals 9

36 divided by a number equals 6

a number times 7 equals 21

$n \times 7 = 21$

$n \times 6 = 30$

$6 \times 7 = n$

$7 \times n = 42$

$36 \div n = 6$

$63 \div n = 9$

2. Write an equation to represent the tape diagram below, and solve for the unknown.

| 8 | 8 | 8 | 8 | 8 | 8 |

k

Equation: _____

Lesson 7: Interpret the unknown in multiplication and division to model and solve problems using units of 6 and 7.

©2018 Great Minds®. eureka-math.org

43

3. Model each problem with a drawing. Then, write an equation using a letter to represent the unknown, and solve for the unknown.

a. Each student gets 3 pencils. There are a total of 21 pencils. How many students are there?

b. Henry spends 24 minutes practicing 6 different basketball drills. He spends the same amount of time on each drill. How much time does Henry spend on each drill?

c. Jessica has 8 pieces of yarn for a project. Each piece of yarn is 6 centimeters long. What is the total length of the yarn?

d. Ginny measures 6 milliliters of water into each beaker. She pours a total of 54 milliliters. How many beakers does Ginny use?

Lesson 7: Interpret the unknown in multiplication and division to model and solve problems using units of 6 and 7.

Name _____   Date _____

Model each problem with a drawing. Then, write an equation using a letter to represent the unknown, and solve for the unknown.

1. Three boys and three girls each buy 7 bookmarks. How many bookmarks do they buy all together?

2. Seven friends equally share the cost of a $56 meal. How much does each person pay?

Lesson 7:   Interpret the unknown in multiplication and division to model and solve problems using units of 6 and 7.

©2018 Great Minds®. eureka-math.org

45

Richard has 2 cartons with 6 eggs in each. As he opens the cartons, he drops 2 eggs. How many unbroken eggs does Richard have left?

_____

_____

_____

_____

Read   **Draw**   **Write**

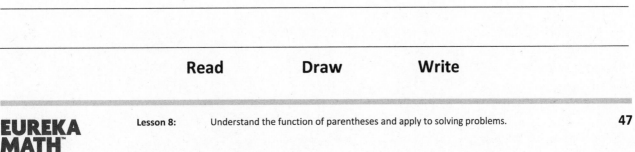

Name _____   Date _____

1. Solve.

   a. $(12 - 4) + 6 = $ _____

   b. $12 - (4 + 6) = $ _____

   c. _____ $= 15 - (7 + 3)$

   d. _____ $= (15 - 7) + 3$

   e. _____ $= (3 + 2) \times 6$

   f. _____ $= 3 + (2 \times 6)$

   g. $4 \times (7 - 2) = $ _____

   h. $(4 \times 7) - 2 = $ _____

   i. _____ $= (12 \div 2) + 4$

   j. _____ $= 12 \div (2 + 4)$

   k. $9 + (15 \div 3) = $ _____

   l. $(9 + 15) \div 3 = $ _____

   m. $60 \div (10 - 4) = $ _____

   n. $(60 \div 10) - 4 = $ _____

   o. _____ $= 35 + (10 \div 5)$

   p. _____ $= (35 + 10) \div 5$

2. Use parentheses to make the equations true.

a. $16 - 4 + 7 = 19$	b. $16 - 4 + 7 = 5$
c. $2 = 22 - 15 + 5$	d. $12 = 22 - 15 + 5$
e. $3 + 7 \times 6 = 60$	f. $3 + 7 \times 6 = 45$
g. $5 = 10 \div 10 \times 5$	h. $50 = 100 \div 10 \times 5$
i. $26 - 5 \div 7 = 3$	j. $36 = 4 \times 25 - 16$

3. The teacher writes 24 ÷ 4 + 2 = _____ on the board. Chad says it equals 8. Samir says it equals 4. Explain how placing the parentheses in the equation can make both answers true.

4. Natasha solves the equation below by finding the sum of 5 and 12. Place the parentheses in the equation to show her thinking. Then, solve.

12 + 15 ÷ 3 = _____

5. Find two possible answers to the expression 7 + 3 × 2 by placing the parentheses in different places.

Lesson 8: Understand the function of parentheses and apply to solving problems.

Name _____   Date _____

1. Use parentheses to make the equations true.

    a. $24 = 32 - 14 + 6$ b. $12 = 32 - 14 + 6$

    c. $2 + 8 \times 7 = 70$ d. $2 + 8 \times 7 = 58$

2. Marcos solves $24 \div 6 + 2 =$ _____. He says it equals 6. Iris says it equals 3. Show how the position of parentheses in the equation can make both answers true.

Name _____   Date _____

Solve the following pairs of problems. Circle the pairs where both problems have the same answer.

1. a. $7 + (6 + 4)$

     b. $(7 + 6) + 4$

5. a. $(3 + 2) \times 5$

     b. $3 + (2 \times 5)$

2. a. $(3 \times 2) \times 4$

     b. $3 \times (2 \times 4)$

6. a. $(8 \div 2) \times 2$

     b. $8 \div (2 \times 2)$

3. a. $(2 \times 1) \times 5$

     b. $2 \times (1 \times 5)$

7. a. $(9 - 5) + 3$

     b. $9 - (5 + 3)$

4. a. $(4 \times 2) \times 2$

     b. $4 \times (2 \times 2)$

8. a. $(8 \times 5) - 4$

     b. $8 \times (5 - 4)$

Name _____   Date _____

1. Use the array to complete the equation.

a. $3 \times 12 =$ _____

b. $(3 \times 3) \times 4$

   = _____ $\times 4$

   = _____

c. $3 \times 14 =$ _____

d. (_____ $\times$ _____) $\times 7$

   = _____ $\times$ _____

   = _____

EUREKA
MATH™

©2018 Great Minds®. eureka-math.org

2. Place parentheses in the equations to simplify. Then, solve. The first one has been done for you.

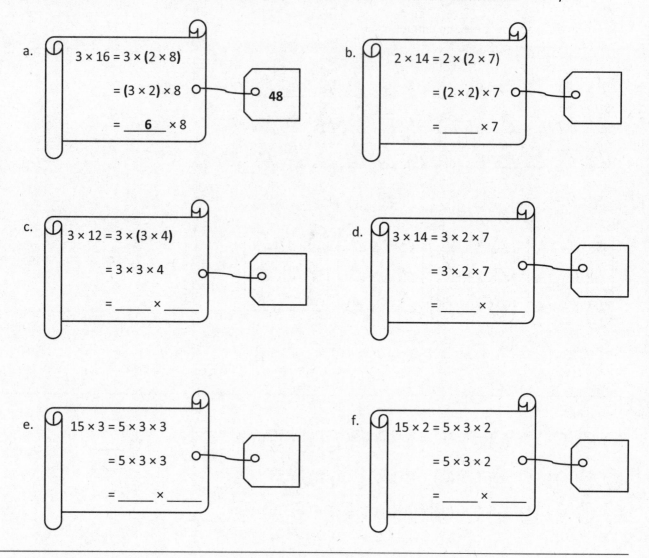

a.
$3 \times 16 = 3 \times (2 \times 8)$
$= (3 \times 2) \times 8$
$= \underline{\quad 6 \quad} \times 8$

48

b.
$2 \times 14 = 2 \times (2 \times 7)$
$= (2 \times 2) \times 7$
$= \underline{\qquad} \times 7$

c.
$3 \times 12 = 3 \times (3 \times 4)$
$= 3 \times 3 \times 4$
$= \underline{\qquad} \times \underline{\qquad}$

d.
$3 \times 14 = 3 \times 2 \times 7$
$= 3 \times 2 \times 7$
$= \underline{\qquad} \times \underline{\qquad}$

e.
$15 \times 3 = 5 \times 3 \times 3$
$= 5 \times 3 \times 3$
$= \underline{\qquad} \times \underline{\qquad}$

f.
$15 \times 2 = 5 \times 3 \times 2$
$= 5 \times 3 \times 2$
$= \underline{\qquad} \times \underline{\qquad}$

3. Charlotte finds the answer to 16×2 by thinking about 8×4. Explain her strategy.

Lesson 9:   Model the associative property as a strategy to multiply.

©2018 Great Minds®. eureka-math.org

Name _____   Date _____

Simplify to find the answer to 18 × 3. Show your work, and explain your strategy.

Lesson 9:   Model the associative property as a strategy to multiply.

57

©2018 Great Minds®. eureka-math.org

Use the 5 plus something break apart and distribute strategy to solve 6 × 8. Model with a tape diagram.

Read   **Draw**   **Write**

Name _____ Date _____

1. Label the arrays. Then, fill in the blanks below to make the statements true.

a. **8 × 8 =** _____

(8 × 5) = _____   ,(8 × _____) = _____

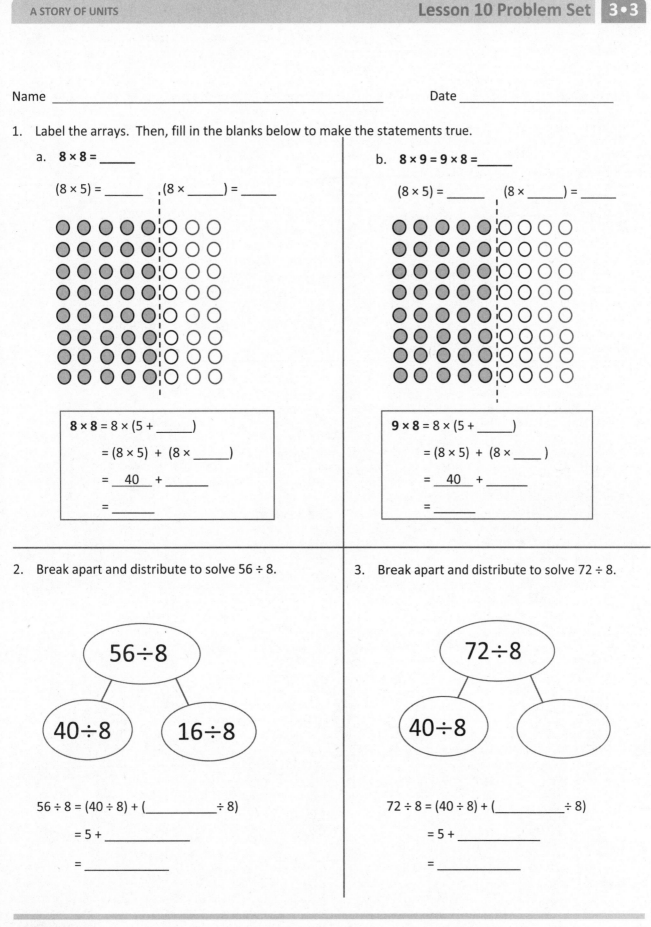

b. **8 × 9 = 9 × 8 =** _____

(8 × 5) = _____   (8 × _____) = _____

8 × 8 = 8 × (5 + _____)

= (8 × 5) + (8 × _____)

= __40__ + _____

= _____

9 × 8 = 8 × (5 + _____)

= (8 × 5) + (8 × _____)

= __40__ + _____

= _____

2. Break apart and distribute to solve 56 ÷ 8.

56÷8

40÷8   16÷8

56 ÷ 8 = (40 ÷ 8) + (_____ ÷ 8)

= 5 + _____

= _____

3. Break apart and distribute to solve 72 ÷ 8.

72÷8

40÷8

72 ÷ 8 = (40 ÷ 8) + (_____ ÷ 8)

= 5 + _____

= _____

EUREKA
MATH™

Lesson 10:   Use the distributive property as a strategy to multiply and divide.

61

©2018 Great Minds®. eureka-math.org

4. An octagon has 8 sides. Skip-count to find the total number of sides on 9 octagons.

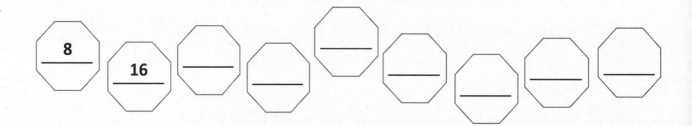

Nine octagons have a total of _____ sides.

5. Multiply.

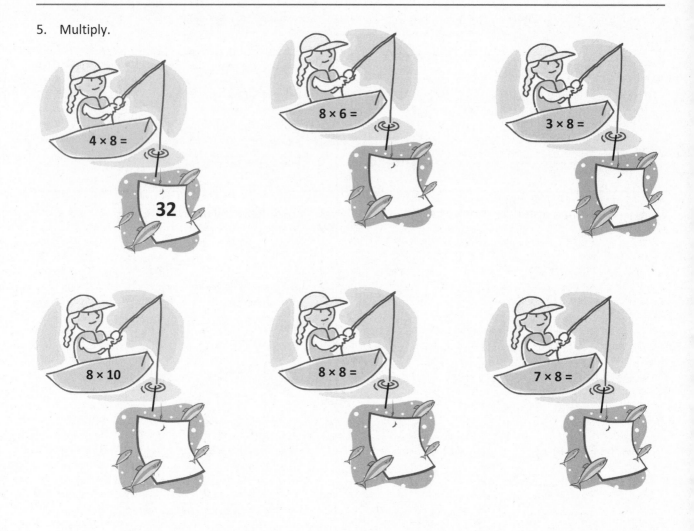

Lesson 10:   Use the distributive property as a strategy to multiply and divide.

©2018 Great Minds®. eureka-math.org

6. Match.

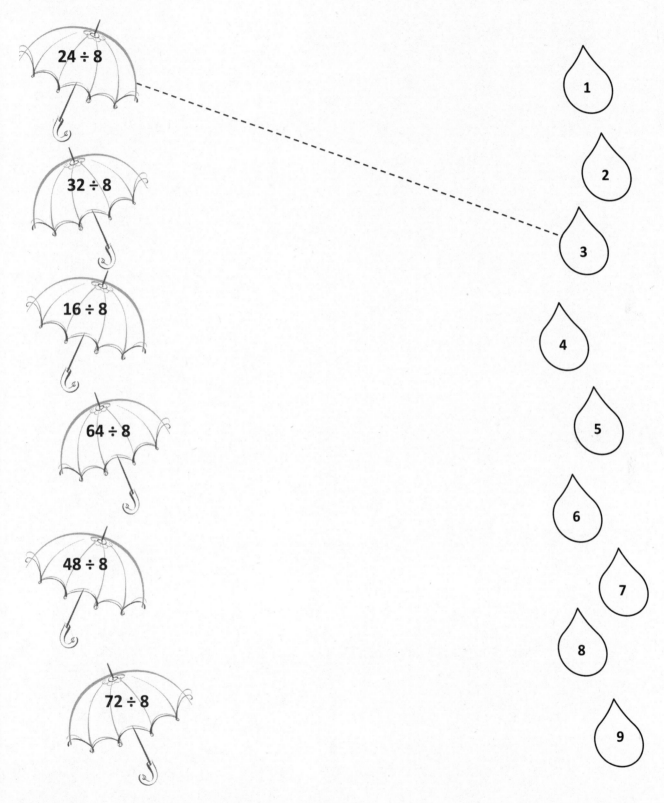

Lesson 10:   Use the distributive property as a strategy to multiply and divide.

63

©2018 Great Minds®. eureka-math.org

Name _____   Date _____

Use the break apart and distribute strategy to solve the following problem. You may choose whether or not to draw an array.

7 × 8 =_____

Name _____   Date _____

1. Ms. Santor divides 32 students into 8 equal groups for a field trip. Draw a tape diagram, and label the number of students in each group as *n*. Write an equation, and solve for *n*.

---

2. Tara buys 6 packs of printer paper. Each pack of paper costs \$8. Draw a tape diagram, and label the total amount she spends as *m*. Write an equation, and solve for *m*.

---

3. Mr. Reed spends \$24 on coffee beans. How many kilograms of coffee beans does he buy? Draw a tape diagram, and label the total amount of coffee beans he buys as *c*. Write an equation, and solve for *c*.

\$8 for 1 kg

Lesson 11:   Interpret the unknown in multiplication and division to model and solve problems.

©2018 Great Minds®. eureka-math.org

67

4. Eight boys equally share 4 packs of baseball cards. Each pack contains 10 cards. How many cards does each boy get?

5. There are 8 bags of yellow and green balloons. Each bag contains 7 balloons. If there are 35 yellow balloons, how many green balloons are there?

6. The fruit seller packs 72 oranges into bags of 8 each. He sells all the oranges at $4 a bag. How much money did he receive?

Lesson 11:   Interpret the unknown in multiplication and division to model and solve problems.

©2018 Great Minds®. eureka-math.org

Name _____   Date _____

Erica buys some packs of rubber bracelets. There are 8 bracelets in each pack.

a. How many packs of rubber bracelets does she buy if she has a total of 56 bracelets? Draw a tape diagram, and label the total number of packages as *p*. Write an equation, and solve for *p*.

b. After giving some bracelets away, Erica has 18 left. How many bracelets did she give away?

EUREKA MATH™

Lesson 11:   Interpret the unknown in multiplication and division to model and
                solve problems.

©2018 Great Minds®. eureka-math.org

69

A scientist fills 5 test tubes with 9 milliliters of fresh water in each. She fills another 3 test tubes with 9 milliliters of salt water in each. How many milliliters of water does she use in all? Use the break apart and distribute strategy to solve.

_____

_____

_____

_____

Read   **Draw**   **Write**

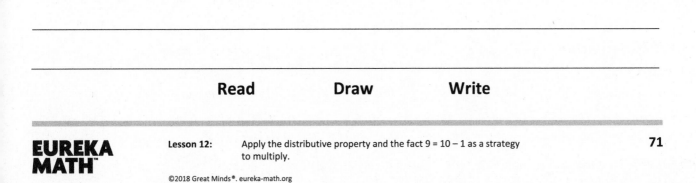

EUREKA
MATH™

Lesson 12:   Apply the distributive property and the fact 9 = 10 − 1 as a strategy
                to multiply.

©2018 Great Minds®. eureka-math.org

71

Name _____   Date _____

1. Each 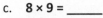 has a value of **9**. Find the value of each row. Then, add the rows to find the total.

a. **6 × 9 =** _____

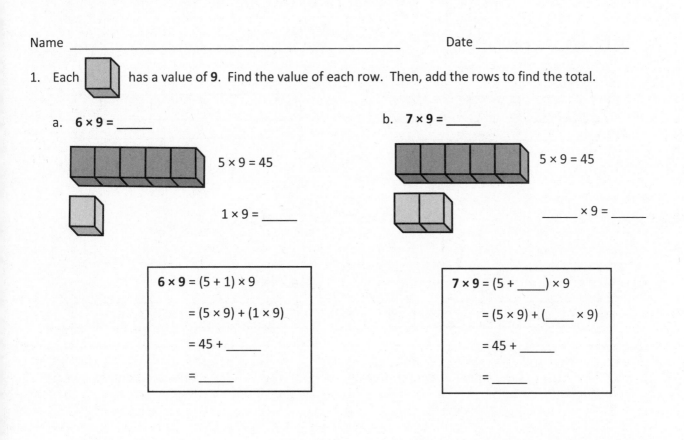

5 × 9 = 45

1 × 9 = _____

$6 \times 9 = (5 + 1) \times 9$
$= (5 \times 9) + (1 \times 9)$
$= 45 + \rule{2em}{0.4pt}$
$= \rule{2em}{0.4pt}$

b. **7 × 9 =** _____

5 × 9 = 45

_____ × 9 = _____

$7 \times 9 = (5 + \rule{2em}{0.4pt}) \times 9$
$= (5 \times 9) + (\rule{2em}{0.4pt} \times 9)$
$= 45 + \rule{2em}{0.4pt}$
$= \rule{2em}{0.4pt}$

c. **8 × 9 =** _____

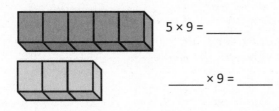

5 × 9 = _____

_____ × 9 = _____

$8 \times 9 = (5 + \rule{2em}{0.4pt}) \times 9$
$= (5 \times 9) + (\rule{2em}{0.4pt} \times \rule{2em}{0.4pt})$
$= 45 + \rule{2em}{0.4pt}$
$= \rule{2em}{0.4pt}$

d. **9 × 9 =** _____

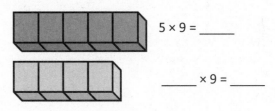

5 × 9 = _____

_____ × 9 = _____

$9 \times 9 = (5 + \rule{2em}{0.4pt}) \times 9$
$= (5 \times 9) + (\rule{2em}{0.4pt} \times \rule{2em}{0.4pt})$
$= 45 + \rule{2em}{0.4pt}$
$= \rule{2em}{0.4pt}$

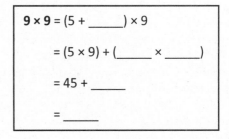

Lesson 12:   Apply the distributive property and the fact 9 = 10 – 1 as a strategy
to multiply.

73

©2018 Great Minds®. eureka-math.org

2. Find the total value of the shaded blocks.

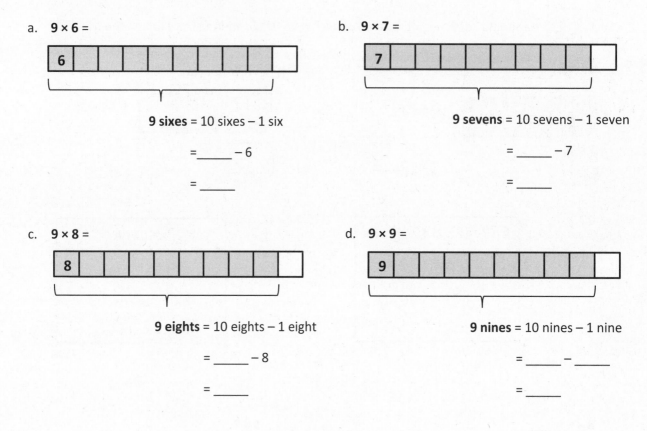

a. **9 × 6 =**

| 6 | | | | | | | | | |

9 **sixes** = 10 sixes − 1 six

= _____ − 6

= _____

b. **9 × 7 =**

| 7 | | | | | | | | | |

9 **sevens** = 10 sevens − 1 seven

= _____ − 7

= _____

c. **9 × 8 =**

| 8 | | | | | | | | | |

9 **eights** = 10 eights − 1 eight

= _____ − 8

= _____

d. **9 × 9 =**

| 9 | | | | | | | | | |

9 **nines** = 10 nines − 1 nine

= _____ − _____

= _____

3. Matt buys a pack of postage stamps. He counts 9 rows of 4 stamps. He thinks of 10 fours to find the total number of stamps. Show the strategy that Matt might have used to find the total number of stamps.

Lesson 12:   Apply the distributive property and the fact 9 = 10 − 1 as a strategy to multiply.

©2018 Great Minds®. eureka-math.org

4. Match.

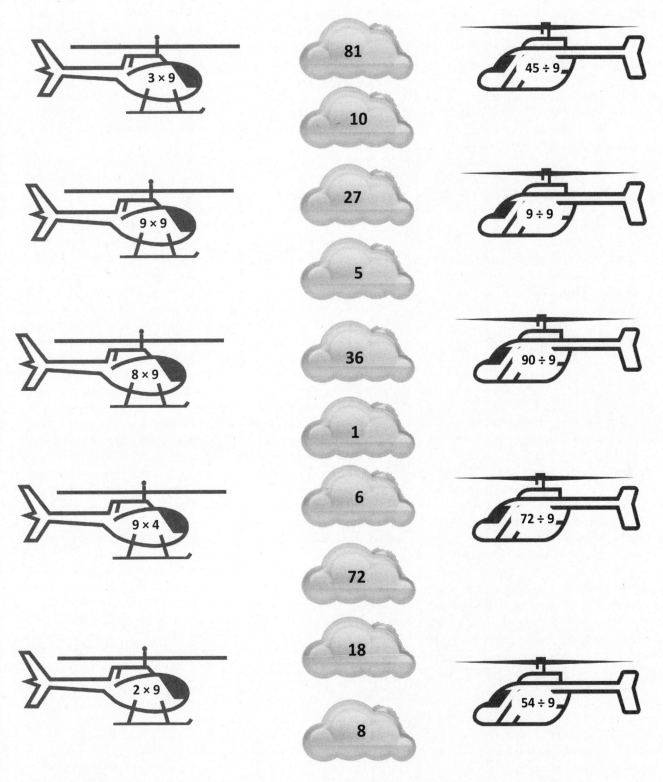

EUREKA MATH

Lesson 12: Apply the distributive property and the fact 9 = 10 − 1 as a strategy to multiply.

©2018 Great Minds®. eureka-math.org

75

Name _____   Date _____

1. Each 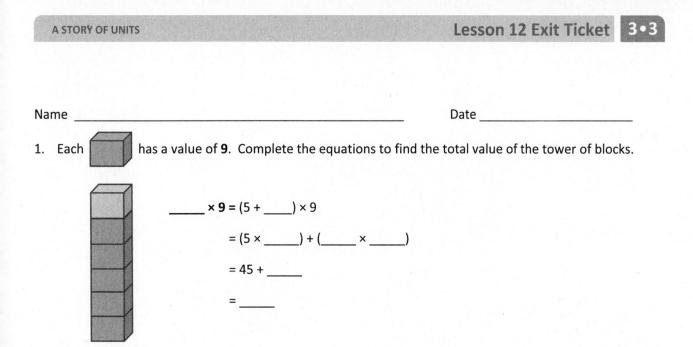 has a value of **9**. Complete the equations to find the total value of the tower of blocks.

_____ × **9** = (5 + _____) × 9

= (5 × _____) + (_____ × _____)

= 45 + _____

= _____

2. Hector solves 9 × 8 by subtracting 1 eight from 10 eights. Draw a model, and explain Hector's strategy.

      Lesson 12:   Apply the distributive property and the fact 9 = 10 − 1 as a strategy   **77**
to multiply.

©2018 Great Minds®. eureka-math.org

tape diagram

Lesson 12:   Apply the distributive property and the fact 9 = 10 − 1 as a strategy
to multiply.

79

Michaela and Gilda read the same book. It takes Michaela about 8 minutes to read a chapter and Gilda about 10 minutes. There are 9 chapters in the book. How many fewer minutes does Michaela spend reading than Gilda?

_____

_____

_____

Read   **Draw**   **Write**

Name _____   Date _____

1. a. Skip-count by nine.

    <u> 9 </u> , _____, _____, _____, _____, _____, _____, _____, _____, _____
                      36   72

    b. Look at the *tens* place in the count-by. What is the pattern?

    c. Look at the *ones* place in the count-by. What is the pattern?

2. Complete to make true statements.

    a. 10 more than 0 is <u> 10 </u> ,

        1 less is <u> 9 </u> .

        $1 \times 9 =$ <u> 9 </u>

    f. 10 more than 45 is _____ ,

        1 less is _____ .

        $6 \times 9 =$ _____

    b. 10 more than 9 is <u> 19 </u> ,

        1 less is <u> 18 </u> .

        $2 \times 9 =$ _____

    g. 10 more than 54 is _____ ,

        1 less is _____ .

        $7 \times 9 =$ _____

    c. 10 more than 18 is _____ ,

        1 less is _____ .

        $3 \times 9 =$ _____

    h. 10 more than 63 is _____ ,

        1 less is _____ .

        $8 \times 9 =$ _____

    d. 10 more than 27 is _____ ,

        1 less is _____ .

        $4 \times 9 =$ _____

    i. 10 more than 72 is _____ ,

        1 less is _____ .

        $9 \times 9 =$ _____

    e. 10 more than 36 is _____ ,

        1 less is _____ .

        $5 \times 9 =$ _____

    j. 10 more than 81 is _____ ,

        1 less is _____ .

        $10 \times 9 =$ _____

3. a. Analyze the equations in Problem 2. What is the pattern?

   b. Use the pattern to find the next 4 facts. Show your work.

   $11 \times 9 =$ $12 \times 9 =$ $13 \times 9 =$ $14 \times 9 =$

   c. Kent notices another pattern in Problem 2. His work is shown below. He sees the following:
      - The tens digit in the product is 1 less than the number of groups.
      - The ones digit in the product is 10 minus the number of groups.

	Tens digit	Ones digit
$2 \times 9 = \underline{18}$ →	$\underline{1} = 2 - 1$	$\underline{8} = 10 - 2$
$3 \times 9 = \underline{27}$ →	$\underline{2} = 3 - 1$	$\underline{7} = 10 - 3$
$4 \times 9 = \underline{36}$ →	$\underline{3} = 4 - 1$	$\underline{6} = 10 - 4$
$5 \times 9 = \underline{45}$ →	$\underline{4} = 5 - 1$	$\underline{5} = 10 - 5$

   Use Kent's strategy to solve 6×9 and 7×9.

   d. Show an example of when Kent's pattern doesn't work.

EUREKA MATH™

4. Each equation contains a letter representing the unknown. Find the value of each unknown. Then, write the letters that match the answers to solve the riddle.

$a \times 9 = 54$

$a =$ _____

$81 \div 9 = g$

$g =$ _____

$9 \times d = 72$

$d =$ _____

$e \times 9 = 63$

$e =$ _____

$o \div 9 = 10$

$o =$ _____

$9 \times n = 27$

$n =$ _____

$t \times 9 = 18$

$t =$ _____

$9 \times s = 36$

$s =$ _____

$i \div 9 = 5$

$i =$ _____

How do you make one vanish?

___ ___ ___   ___ " ___ "   ___ ___ ___   ___ ___ ___ ,   ___ ___ ___ ___ !
 6   8   8   6   9   6   3   8   45   2   4   9   90   3   7

Name _____   Date _____

1. $6 \times 9 = 54$ $8 \times 9 = 72$

    What is 10 more than 54? _____   What is 10 more than 72? _____

    What is 1 less? _____   What is 1 less? _____

    $7 \times 9 =$ _____ $9 \times 9 =$ _____

2. Explain the pattern used in Problem 1.

Name _____   Date _____

1. a. Multiply. Then, add the tens digit and ones digit of each product.

1 × 9 = 9	__0__ + __9__ = __9__
2 × 9 = 18	__1__ + __8__ = _____
3 × 9 =	_____ + _____ = _____
4 × 9 =	_____ + _____ = _____
5 × 9 =	_____ + _____ = _____
6 × 9 =	_____ + _____ = _____
7 × 9 =	_____ + _____ = _____
8 × 9 =	_____ + _____ = _____
9 × 9 =	_____ + _____ = _____
10 × 9 =	_____ + _____ = _____

b. What is the sum of the digits in each product? How can this strategy help you check your work with the nines facts?

c. Araceli continues to count by nines. She writes, "90, 99, 108, 117, 126, 135, 144, 153, 162, 171, 180, 189, 198. Wow! The sum of the digits is still 9." Is she correct? Why or why not?

2. Araceli uses the number of groups in 8 × 9 to help her find the product. She uses 8 – 1 = 7 to get the digit in the tens place and 10 – 8 = 2 to get the digit in the ones place. Use her strategy to find 4 more facts.

3. Dennis calculates 9 × 8 by thinking about it as 80 – 8 = 72. Explain Dennis' strategy.

4. Sonya figures out the answer to 7 × 9 by putting down her right index finger (shown). What is the answer? Explain how to use Sonya's finger strategy.

EUREKA MATH™

Name _____   Date _____

Donald writes 6 × 9 = 54. Explain two strategies you could use to check his work.

Name _____ Date _____

Write an equation, and use a letter to represent the unknown for Problems 1–6.

1. Mrs. Parson gave each of her grandchildren $9. She gave a total of $36. How many grandchildren does Mrs. Parson have?

2. Shiva pours 27 liters of water equally into 9 containers. How many liters of water are in each container?

3. Derek cuts 7 pieces of wire. Each piece is 9 meters long. What is the total length of the 7 pieces?

Lesson 15: Interpret the unknown in multiplication and division to model and solve problems.

©2018 Great Minds®. eureka-math.org

93

4. Aunt Deena and Uncle Chris share the cost of a limousine ride with their 7 friends. The ride cost a total of $63. If everyone shares the cost equally, how much does each person pay?

5. Cara bought 9 packs of beads. There are 10 beads in each pack. She always uses 30 beads to make each necklace. How many necklaces can she make if she uses all the beads?

6. There are 8 erasers in a set. Damon buys 9 sets. After giving some erasers away, Damon has 35 erasers left. How many erasers did he give away?

Lesson 15: Interpret the unknown in multiplication and division to model and solve problems.

Name _____   Date _____

Use a letter to represent the unknown.

1. Mrs. Aquino pours 36 liters of water equally into 9 containers. How much water is in each container?

2. Marlon buys 9 packs of hot dogs. There are 6 hot dogs in each pack. After the barbeque, 35 hot dogs are left over. How many hot dogs were eaten?

Lesson 15: Interpret the unknown in multiplication and division to model and solve problems.

©2018 Great Minds®. eureka-math.org

95

Name _____   Date _____

1. Complete.

    a. _____ × 1 = 6   b. _____ ÷ 7 = 0   c. 8 × _____ = 8   d. 9 ÷ _____ = 9

    e. 0 ÷ 5 = _____   f. _____ × 0 = 0   g. 4 ÷ _____ = 1   h. _____ × 1 = 3

2. Match each equation with its solution.

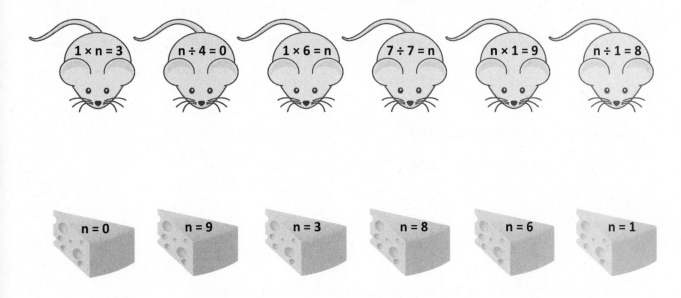

3. Let *n* be a number. Complete the blanks below with the products.

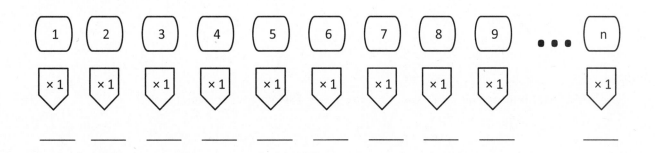

    What pattern do you notice?

Lesson 16:   Reason about and explain arithmetic patterns using units   **97**
of 0 and 1 as they relate to multiplication and division.

©2018 Great Minds®. eureka-math.org

4. Josie says that any number divided by 1 equals that number.

   a. Write a division equation using *n* to represent Josie's statement.

   b. Use your equation from Part (a). Let *n* = 6. Write a new equation, and draw a picture to show that your equation is true.

   c. Write the related multiplication equation that you can use to check your division equation.

5. Matt explains what he learned about dividing with zero to his little sister.

   a. What might Matt tell his sister about solving $0 \div 9$? Explain your answer.

   b. What might Matt tell his sister about solving $8 \div 0$? Explain your answer.

   c. What might Matt tell his sister about solving $0 \div 0$? Explain your answer.

Lesson 16:   Reason about and explain arithmetic patterns using units of 0 and 1 as they relate to multiplication and division.

EUREKA MATH™

Name _____   Date _____

1. Complete.

   a. _____ × 1 = 5   b. 6 × _____ = 6   c. _____ ÷ 7 = 0

   d. 5 × _____ = 0   e. 1 = 9 ÷ _____   f. 8 = 1 × _____

2. Luis divides 8 by 0 and says it equals 0. Is he correct? Explain why or why not.

Lesson 16:   Reason about and explain arithmetic patterns using units
of 0 and 1 as they relate to multiplication and division.

©2018 Great Minds®. eureka-math.org

99

Henry's garden has 9 rows of squash plants. Each row has 8 squash plants. There is also 1 row with 8 watermelon plants. How many squash and watermelon plants does Henry have in all?

_____

_____

_____

_____

Read   **Draw**   **Write**

EUREKA MATH™

Lesson 17: Identify patterns in multiplication and division facts using the multiplication table.

101

©2018 Great Minds®. eureka-math.org

Name _____   Date _____

1. Write the products into the squares as fast as you can.

1 × 1	2 × 1	3 × 1	4 × 1	5 × 1	6 × 1	7 × 1	8 × 1
1 × 2	2 × 2	3 × 2	4 × 2	5 × 2	6 × 2	7 × 2	8 × 2
1 × 3	2 × 3	3 × 3	4 × 3	5 × 3	6 × 3	7 × 3	8 × 3
1 × 4	2 × 4	3 × 4	4 × 4	5 × 4	6 × 4	7 × 4	8 × 4
1 × 5	2 × 5	3 × 5	4 × 5	5 × 5	6 × 5	7 × 5	8 × 5
1 × 6	2 × 6	3 × 6	4 × 6	5 × 6	6 × 6	7 × 6	8 × 6
1 × 7	2 × 7	3 × 7	4 × 7	5 × 7	6 × 7	7 × 7	8 × 7
1 × 8	2 × 8	3 × 8	4 × 8	5 × 8	6 × 8	7 × 8	8 × 8

a. Color all the squares with even products orange. Can an even product ever have an odd factor?

b. Can an odd product ever have an even factor?

c. Everyone knows that 7 × 4 = (5 × 4) + (2 × 4). Explain how this is shown in the table.

d. Use what you know to find the product of 7 × 16 or 8 sevens + 8 sevens.

EUREKA MATH™

Lesson 17:   Identify patterns in multiplication and division facts using the multiplication table.

103

©2018 Great Minds®. eureka-math.org

2. In the table, only the products on the diagonal are shown.

   a. Label each product on the diagonal.

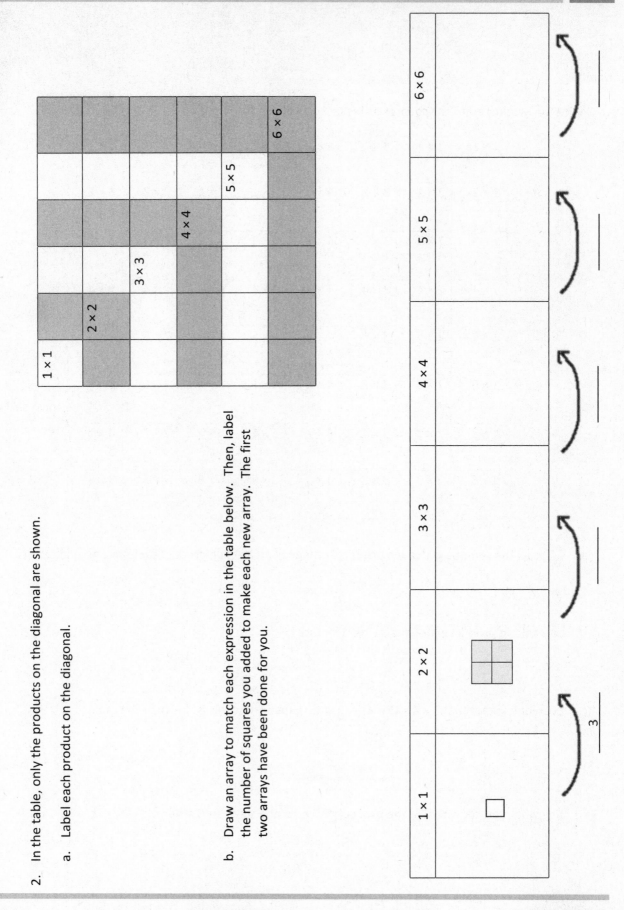

   b. Draw an array to match each expression in the table below. Then, label the number of squares you added to make each new array. The first two arrays have been done for you.

Lesson 17: Identify patterns in multiplication and division facts using the multiplication table.

EUREKA MATH

c. What pattern do you notice in the number of squares that are added to each new array?

d. Use the pattern you discovered in Part (b) to prove this: 9 × 9 is the sum of the first 9 odd numbers.

Lesson 17:   Identify patterns in multiplication and division facts using the
             multiplication table.

©2018 Great Minds®. eureka-math.org

105

Name _____   Date _____

1. Use what you know to find the product of 8 × 12 or 6 eights + 6 eights.

2. Luis says 3 × 233 = 626. Use what you learned about odd times odd to explain why Luis is wrong.

Lesson 17:   Identify patterns in multiplication and division facts using the
                    multiplication table.

©2018 Great Minds®. eureka-math.org

107

Name _____ Date _____

Use the RDW process for each problem. Explain why your answer is reasonable.

1. Rose has 6 pieces of yarn that are each 9 centimeters long. Sasha gives Rose a piece of yarn. Now, Rose has a total of 81 centimeters of yarn. What is the length of the yarn that Sasha gives Rose?

2. Julio spends 29 minutes doing his spelling homework. He then completes each math problem in 4 minutes. There are 7 math problems. How many minutes does Julio spend on his homework in all?

Lesson 18:   Solve two-step word problems involving all four operations and assess the reasonableness of solutions.

©2018 Great Minds®. eureka-math.org

109

3. Pearl buys 125 stickers. She gives 53 stickers to her little sister. Pearl then puts 9 stickers on each page of her album. If she uses all of her remaining stickers, on how many pages does Pearl put stickers?

4. Tanner's beaker had 45 milliliters of water in it at first. After each of his friends poured in 8 milliliters, the beaker contained 93 milliliters. How many friends poured water into Tanner's beaker?

5. Cora weighs 4 new, identical pencils and a ruler. The total weight of these items is 55 grams. She weighs the ruler by itself and it weighs 19 grams. How much does each pencil weigh?

Lesson 18:   Solve two-step word problems involving all four operations and assess the reasonableness of solutions.

©2018 Great Minds®. eureka-math.org

Name _____　Date _____

Use the RDW process to solve. Explain why your answer is reasonable.

On Saturday, Warren swims laps in the pool for 45 minutes. On Sunday, he runs 8 miles. It takes him 9 minutes to run each mile. How long does Warren spend exercising over the weekend?

Lesson 18:　　Solve two-step word problems involving all four operations and assess the reasonableness of solutions.　　　　　　　　　111

©2018 Great Minds®. eureka-math.org

Mia has 152 beads. She uses some to make bracelets. Now there are 80 beads. If she uses 8 beads for each bracelet, how many bracelets does she make?

_____

_____

_____

_____

Read   **Draw**   **Write**

Lesson 19:   Multiply by multiples of 10 using the place value chart.

113

©2018 Great Minds®. eureka-math.org

Name _____   Date _____

1. Use the disks to fill in the blanks in the equations.

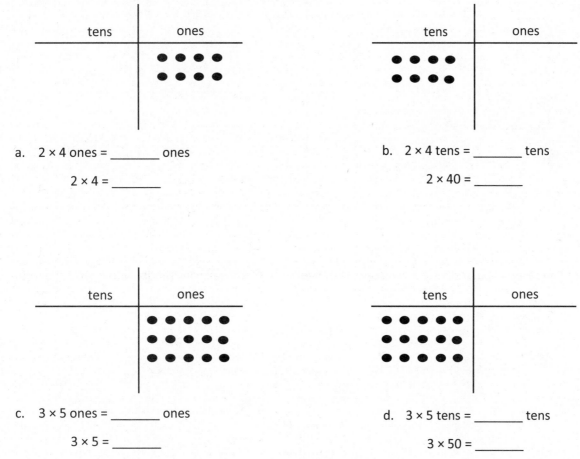

a.

4 × 3 ones = _____ ones

4 × 3 = _____

b.

4 × 3 tens = _____ tens

4 × 30 = _____

2. Use the chart to complete the blanks in the equations.

tens	ones
	● ● ● ●
	● ● ● ●

a. 2 × 4 ones = _____ ones

2 × 4 = _____

tens	ones
● ● ● ●	
● ● ● ●	

b. 2 × 4 tens = _____ tens

2 × 40 = _____

tens	ones
	● ● ● ● ●
	● ● ● ● ●
	● ● ● ● ●

c. 3 × 5 ones = _____ ones

3 × 5 = _____

tens	ones
● ● ● ● ●	
● ● ● ● ●	
● ● ● ● ●	

d. 3 × 5 tens = _____ tens

3 × 50 = _____

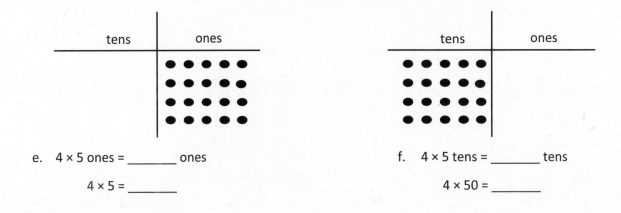

e.   4 × 5 ones = _____ ones

    4 × 5 = _____

f.   4 × 5 tens = _____ tens

    4 × 50 = _____

3.   Fill in the blank to make the equation true.

a. _____ = 7 × 2	b. _____ tens = 7 tens × 2
c. _____ = 8 × 3	d. _____ tens = 8 tens × 3
e. _____ = 60 × 5	f. _____ = 4 × 80
g. 7 × 40 = _____	h. 50 × 8 = _____

4.   A bus can carry 40 passengers. How many passengers can 6 buses carry? Model with a tape diagram.

Name _____ Date _____

1. Use the chart to complete the blanks in the equations.

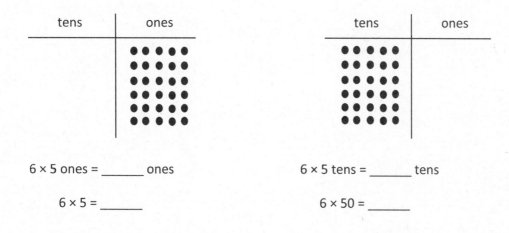

6×5 ones = _____ ones

6×5 = _____

6×5 tens = _____ tens

6×50 = _____

2. A small plane has 20 rows of seats. Each row has 4 seats.

a. Find the total number of seats on the plane.

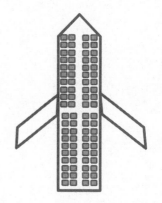

b. How many seats are on 3 small planes?

Model 3 × 4 on a place value chart. Then, explain how the array can help you solve 30 × 4.

_____

_____

_____

_____

Read   **Draw**   **Write**

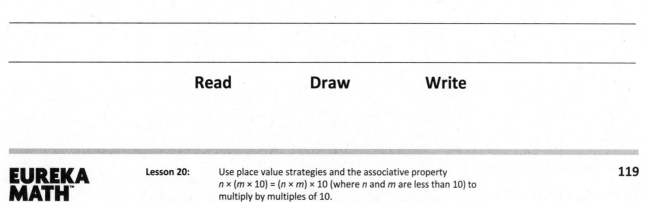

EUREKA MATH

Lesson 20:   Use place value strategies and the associative property
n × (m × 10) = (n × m) × 10 (where n and m are less than 10) to
multiply by multiples of 10.

©2018 Great Minds®. eureka-math.org

119

Name _____   Date _____

1. Use the chart to complete the equations. Then, solve. The first one has been done for you.

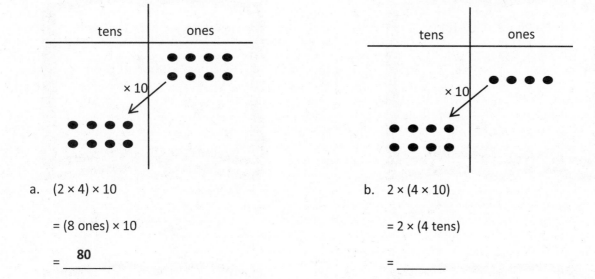

a. (2 × 4) × 10

   = (8 ones) × 10

   = __**80**__

b. 2 × (4 × 10)

   = 2 × (4 tens)

   = _____

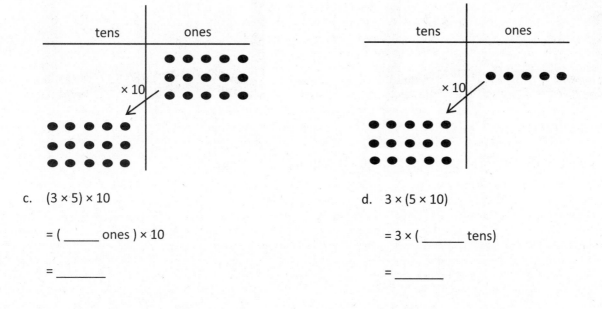

c. (3 × 5) × 10

   = (_____ ones) × 10

   = _____

d. 3 × (5 × 10)

   = 3 × (_____ tens)

   = _____

Lesson 20:   Use place value strategies and the associative property
             n × (m × 10) = (n × m) × 10 (where n and m are less than 10) to
             multiply by multiples of 10.

©2018 Great Minds®. eureka-math.org

121

2. Place parentheses in the equations to find the related fact. Then, solve. The first one has been done for you.

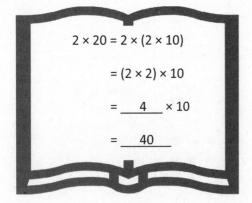

$2 \times 20 = 2 \times (2 \times 10)$

$= (2 \times 2) \times 10$

$= \underline{\quad 4 \quad} \times 10$

$= \underline{\quad 40 \quad}$

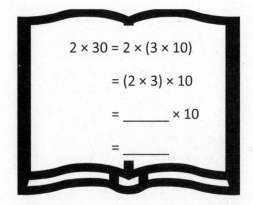

$2 \times 30 = 2 \times (3 \times 10)$

$= (2 \times 3) \times 10$

$= \underline{\qquad} \times 10$

$= \underline{\qquad}$

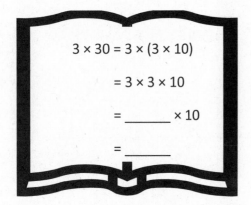

$3 \times 30 = 3 \times (3 \times 10)$

$= 3 \times 3 \times 10$

$= \underline{\qquad} \times 10$

$= \underline{\qquad}$

$2 \times 50 = 2 \times 5 \times 10$

$= 2 \times 5 \times 10$

$= \underline{\qquad} \times 10$

$= \underline{\qquad}$

3. Gabriella solves 20×4 by thinking about 10×8. Explain her strategy.

Lesson 20:   Use place value strategies and the associative property
$n \times (m \times 10) = (n \times m) \times 10$ (where n and m are less than 10) to
multiply by multiples of 10.
©2018 Great Minds®. eureka-math.org

EUREKA
MATH™

Name _____   Date _____

1. Place parentheses in the equations to find the related fact. Then, solve.

    a.   4 × 20 = 4 × 2 × 10   b.   3 × 30 = 3 × 3 × 10

         = 4 × 2 × 10   = 3 × 3 × 10

         = _____ × 10   = _____ × 10

         = _____   = _____

2. Jamila solves 20 × 5 by thinking about 10 tens. Explain her strategy.

Lesson 20:   Use place value strategies and the associative property
                   n × (m × 10) = (n × m) × 10 (where n and m are less than 10) to
                   multiply by multiples of 10.

Name _____ Date _____

Use the RDW process to solve each problem. Use a letter to represent the unknown.

1. There are 60 seconds in 1 minute. Use a tape diagram to find the total number of seconds in 5 minutes and 45 seconds.

2. Lupe saves $30 each month for 4 months. Does she have enough money to buy the art supplies below? Explain why or why not.

Art Supplies
$142

3. Brad receives 5 cents for each can or bottle he recycles. How many cents does Brad earn if he recycles 48 cans and 32 bottles?

EUREKA MATH™

Lesson 21: Solve two-step word problems involving multiplying single-digit factors and multiples of 10.

125

©2018 Great Minds®. eureka-math.org

4. A box of 10 markers weighs 105 grams. If the empty box weighs 15 grams, how much does each marker weigh?

5. Mr. Perez buys 3 sets of cards. Each set comes with 18 striped cards and 12 polka dot cards. He uses 49 cards. How many cards does he have left?

6. Ezra earns $9 an hour working at a book store. She works for 7 hours each day on Mondays and Wednesdays. How much does Ezra earn each week?

Name _____   Date _____

Use the RDW process to solve. Use a letter to represent the unknown.

Frederick buys a can of 3 tennis balls. The empty can weighs 20 grams, and each tennis ball weighs 60 grams. What is the total weight of the can with 3 tennis balls?

Lesson 21:   Solve two-step word problems involving multiplying single-digit factors and multiples of 10.

©2018 Great Minds®. eureka-math.org

127

Grade 3
Module 4

Eric makes a shape with 8 trapezoid pattern blocks. Brock makes the same shape using triangle pattern blocks. It takes 3 triangles to make 1 trapezoid. How many triangle pattern blocks does Brock use?

_____

_____

_____

_____

Read   Draw   Write

EUREKA MATH™

Lesson 1:   Understand area as an attribute of plane figures.

131

©2018 Great Minds®. eureka-math.org

Name _____   Date _____

1. Use triangle pattern blocks to cover each shape below. Draw lines to show where the triangles meet.
    Then, write how many triangle pattern blocks it takes to cover each shape.

    Shape A: _____ triangles

    Shape B: _____ triangles

2. Use rhombus pattern blocks to cover each shape below. Draw lines to show where the rhombuses meet.
    Then, write how many rhombus pattern blocks it takes to cover each shape.

    Shape A: _____ rhombuses

    Shape B: _____ rhombuses

3. Use trapezoid pattern blocks to cover each shape below. Draw lines to show where the trapezoids meet.
    Then, write how many trapezoid pattern blocks it requires to cover each shape.

    Shape A: _____ trapezoids

    Shape B: _____ trapezoids

4. How is the number of pattern blocks needed to cover the same shape related to the size of the pattern blocks?

5. Use square pattern blocks to cover the rectangle below. Draw lines to show where the squares meet. Then, write how many square pattern blocks it requires to cover the rectangle.

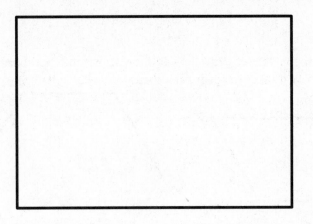

_____ squares

6. Use trapezoid pattern blocks to cover the rectangle in Problem 5. Can you use trapezoid pattern blocks to measure the area of this rectangle? Explain your answer.

Name _____   Date _____

Each [ ] is 1 square unit. Do both rectangles have the same area? Explain how you know.

Wilma and Freddie use pattern blocks to make shapes as shown. Freddie says his shape has a bigger area than Wilma's because it is longer than hers. Is he right? Explain your answer.

Wilma's Shape

Freddie's Shape

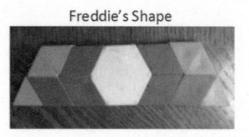

_____

_____

_____

_____

Read   Draw   Write

Name _____   Date _____

1. Use all of Paper Strip 1, which you cut into 12 square inches, to complete the chart below.

	Drawing	Area
Rectangle A		
Rectangle B		
Rectangle C		

2. Use all of Paper Strip 2, which you cut into 12 square centimeters, to complete the chart below.

	Drawing	Area
Rectangle A		
Rectangle B		
Rectangle C		

3. Compare the areas of the rectangles you made with Paper Strip 1 and Paper Strip 2. What changed? Why did it change?

4. Maggie uses square units to create these two rectangles. Do the two rectangles have the same area? How do you know?

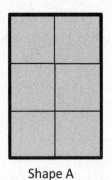

Shape A

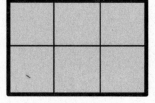

Shape B

5. Count to find the area of the rectangle below. Then, draw a different rectangle that has the same area.

Lesson 2: Decompose and recompose shapes to compare areas.

Name _____   Date _____

1. Each [ ] is a square unit. Find the area of the rectangle below. Then, draw a different rectangle with the same number of square units.

2. Zach creates a rectangle with an area of 6 square inches. Luke makes a rectangle with an area of 6 square centimeters. Do the two rectangles have the same area? Why or why not?

Jace uses paper squares to create a rectangle. Clary cuts all of Jace's squares in half to create triangles. She uses all the triangles to make a rectangle. There are 16 triangles in Clary's rectangle. How many squares were in Jace's shape?

_____

_____

_____

_____

Read   Draw   Write

Lesson 3:   Model tiling with centimeter and inch unit squares as a strategy to
                measure area.

143

©2018 Great Minds®. eureka-math.org

Name _____   Date _____

1. Each ▢ is 1 square unit. What is the area of each of the following rectangles?

A: _____ square units

B: _____

C: _____

D: _____

2. Each ▨ is 1 square unit. What is the area of each of the following rectangles?

a. _____

b. _____

c. _____

d. _____

EUREKA
MATH™

Lesson 3:   Model tiling with centimeter and inch unit squares as a strategy to
              measure area.

©2018 Great Minds®. eureka-math.org

145

3. a. How would the rectangles in Problem 1 be different if they were composed of square inches?

    b. Select one rectangle from Problem 1 and recreate it on square inch and square centimeter grid paper.

4. Use a separate piece of square centimeter grid paper. Draw four different rectangles that each has an area of 8 square centimeters.

Lesson 3: Model tiling with centimeter and inch unit squares as a strategy to measure area.

EUREKA
MATH

Name _____ Date _____

1. Each [ ] is 1 square unit. Write the area of Rectangle A. Then, draw a different rectangle with the same area in the space provided.

Area = _____

2. Each [ ] is 1 square unit. Does this rectangle have the same area as Rectangle A? Explain.

Lesson 3: Model tiling with centimeter and inch unit squares as a strategy to measure area.

©2018 Great Minds®. eureka-math.org

147

centimeter grid

Lesson 3:   Model tiling with centimeter and inch unit squares as a strategy to measure area.

149

©2018 Great Minds®. eureka-math.org

inch grid

Lesson 3:   Model tiling with centimeter and inch unit squares as a strategy to measure area.

©2018 Great Minds®. eureka-math.org

151

Mara uses 15 square-centimeter tiles to make a rectangle. Ashton uses 9 square-centimeter tiles to make a rectangle.

a. Draw what Mara and Ashton's rectangles might look like.

b. Whose rectangle has a bigger area? How do you know?

Read   **Draw**   **Write**

Lesson 4:   Relate side lengths with the number of tiles on a side.

©2018 Great Minds®. eureka-math.org

153

Name _____   Date _____

1. Use a ruler to measure the side lengths of the rectangle in centimeters. Mark each centimeter with a point and connect the points to show the square units. Then, count the squares you drew to find the total area.

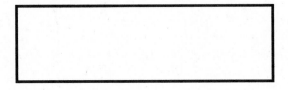

                        Total area: _____

2. Use a ruler to measure the side lengths of the rectangle in inches. Mark each inch with a point and connect the points to show the square units. Then, count the squares you drew to find the total area.

                        Total area: _____

3. Mariana uses square centimeter tiles to find the side lengths of the rectangle below. Label each side length. Then, count the tiles to find the total area.

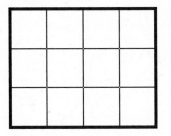

                        Total area: _____

4. Each ▢ is 1 square centimeter. Saffron says that the side length of the rectangle below is

    4 centimeters. Kevin says the side length is 5 centimeters. Who is correct? Explain how you know.

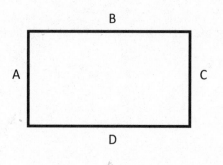

5. Use both square centimeter and square inch tiles to find the area of the rectangle below. Which works best? Explain why.

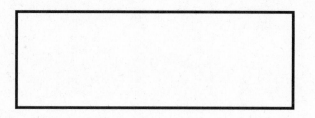

6. How does knowing side lengths A and B help you find side lengths C and D on the rectangle below?

            B

A   C

            D

EUREKA MATH™

Name _____ Date _____

Label the side lengths of each rectangle. Then, match the rectangle to its total area.

a.

12 square centimeters

b.

5 square inches

c.

6 square centimeters

Lesson 4:   Relate side lengths with the number of tiles on a side.

157

©2018 Great Minds®. eureka-math.org

Candice uses square centimeter tiles to find the side lengths of a rectangle as shown on the right. She says the side lengths are 5 centimeters and 7 centimeters. Her partner, Luis, uses a ruler to check Candice's work and says that the side lengths are 5 centimeters and 6 centimeters. Who is right? How do you know?

_____

_____

_____

_____

Read   Draw   Write

EUREKA MATH

Lesson 5:   Form rectangles by tiling with unit squares to make arrays.

159

©2018 Great Minds®. eureka-math.org

Name _____   Date _____

1. Use the centimeter side of a ruler to draw in the tiles. Find the unknown side length or skip-count to find the unknown area. Then, complete the multiplication equations.

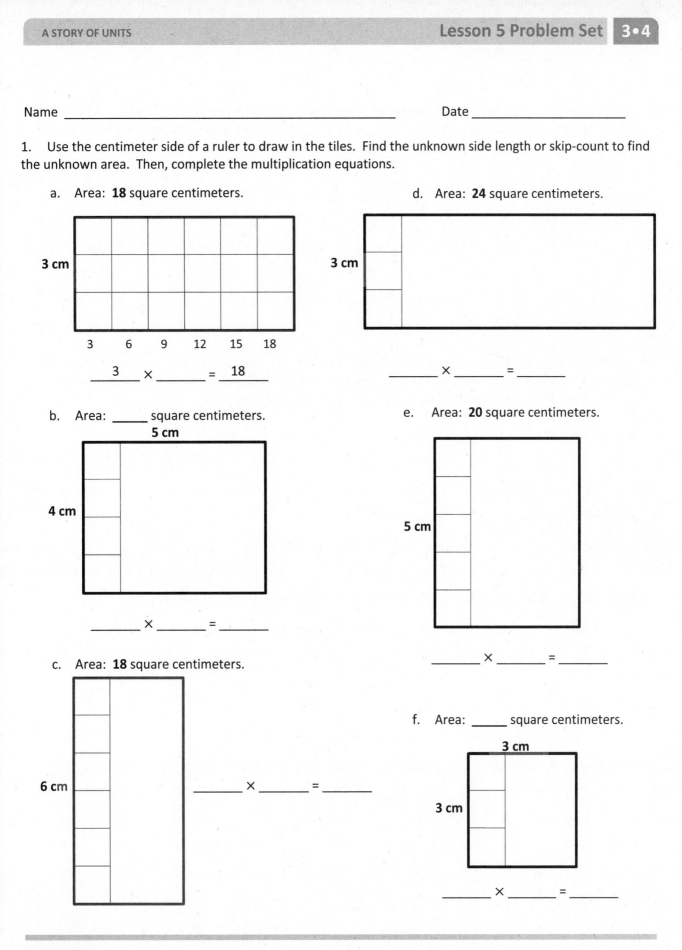

a. Area: **18** square centimeters.

3 cm

3   6   9   12   15   18

___3___ × _____ = ___18___

b. Area: _____ square centimeters.

5 cm

4 cm

_____ × _____ = _____

c. Area: **18** square centimeters.

6 cm

_____ × _____ = _____

d. Area: **24** square centimeters.

3 cm

_____ × _____ = _____

e. Area: **20** square centimeters.

5 cm

_____ × _____ = _____

f. Area: _____ square centimeters.

3 cm

3 cm

_____ × _____ = _____

2. Lindsey makes a rectangle with 35 square inch tiles. She arranges the tiles in 5 equal rows. What are the side lengths of the rectangle? Use words, pictures, and numbers to support your answer.

3. Mark has a total of 24 square inch tiles. He uses 18 square inch tiles to build one rectangular array. He uses the remaining square inch tiles to build a second rectangular array. Draw two arrays that Mark might have made. Then, write multiplication sentences for each.

4. Leon makes a rectangle with 32 square centimeter tiles. There are 4 equal rows of tiles.

    a. How many tiles are in each row? Use words, pictures, and numbers to support your answer.

    b. Can Leon arrange all of his 32 square centimeter tiles into 6 equal rows? Explain your answer.

Lesson 5:   Form rectangles by tiling with unit squares to make arrays.

Name _____   Date _____

Darren has a total of 28 square centimeter tiles. He arranges them into 7 equal rows. Draw Darren's rectangle. Label the side lengths, and write a multiplication sentence to find the total area.

EUREKA MATH

Lesson 5:   Form rectangles by tiling with unit squares to make arrays.

©2018 Great Minds®. eureka-math.org

163

Huma has 4 bags of square inch tiles with 6 tiles in each bag. She uses them to measure the area of a rectangle on her homework. After covering the rectangle, Huma has 4 tiles left. What is the area of the rectangle?

_____

_____

_____

Read   **Draw**   **Write**

EUREKA
MATH™

Lesson 6:   Draw rows and columns to determine the area of a rectangle given an
              incomplete array.

165

©2018 Great Minds®. eureka-math.org

Name _____   Date _____

1. Each □ represents 1 square centimeter. Draw to find the number of rows and columns in each array. Match it to its completed array. Then, fill in the blanks to make a true equation to find each array's area.

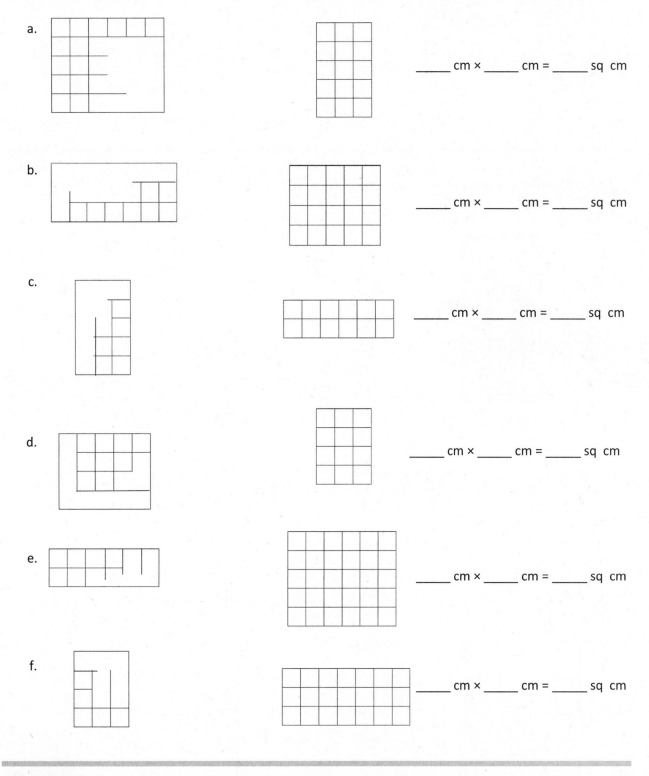

a. _____ cm × _____ cm = _____ sq cm

b. _____ cm × _____ cm = _____ sq cm

c. _____ cm × _____ cm = _____ sq cm

d. _____ cm × _____ cm = _____ sq cm

e. _____ cm × _____ cm = _____ sq cm

f. _____ cm × _____ cm = _____ sq cm

Lesson 6:   Draw rows and columns to determine the area of a rectangle given an incomplete array.

167

2. Sheena skip-counts by sixes to find the total square units in the rectangle below. She says there are 42 square units. Is she right? Explain your answer.

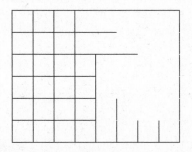

3. The tile floor in Brandon's living room has a rug on it as shown below. How many square tiles are on the floor, including the tiles under the rug?

4. Abdul is creating a stained glass window with square inch glass tiles as shown below. How many more square inch glass tiles does Abdul need to finish his glass window? Explain your answer.

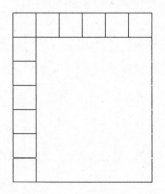

Lesson 6:   Draw rows and columns to determine the area of a rectangle given an incomplete array.

Name _____   Date _____

The tiled floor in Cayden's dining room has a rug on it as shown below. How many square tiles are on the floor, including the tiles under the rug?

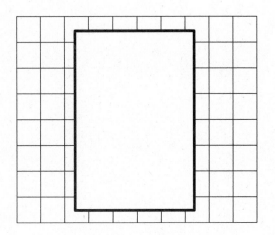

Lesson 6:   Draw rows and columns to determine the area of a rectangle given an incomplete array.

169

©2018 Great Minds®. eureka-math.org

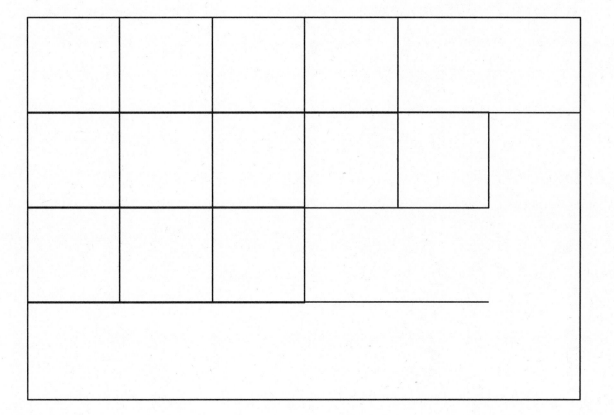

array 1

Lesson 6:   Draw rows and columns to determine the area of a rectangle given an incomplete array.

171

©2018 Great Minds®. eureka-math.org

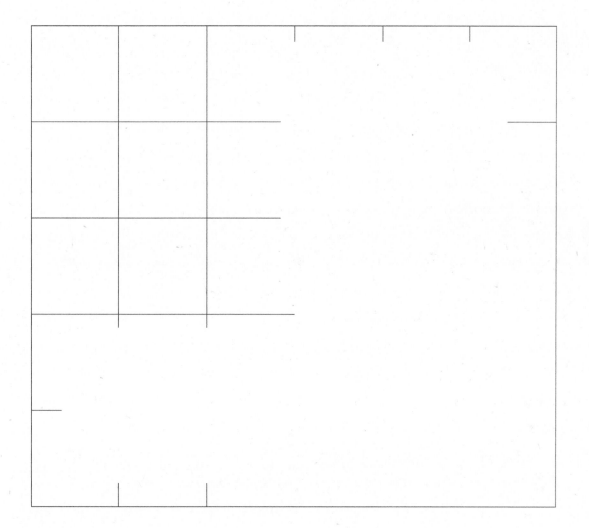

array 2

Lesson 6:   Draw rows and columns to determine the area of a rectangle given an
incomplete array.

173

©2018 Great Minds®. eureka-math.org

Lori wants to replace the square tiles on her wall. The square tiles are sold in boxes of 8 square tiles. Lori buys 6 boxes of tiles. Does she have enough to replace all of the tiles, including the tiles under the painting? Explain your answer.

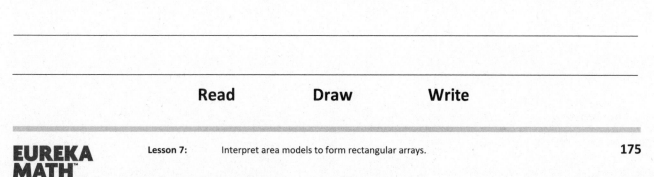

Read   **Draw**   **Write**

Name _____ Date _____

1. Use a straight edge to draw a grid of equal size squares within the rectangle. Find and label the side lengths. Then, multiply the side lengths to find the area.

a. Area A:

   ____ units × ____ units = ____ square units

b. Area B:

   ____ units × ____ units = ____ square units

c. Area C:

   ____ units × ____ units = ____ square units

d. Area D:

   ____ units × ____ units = ____ square units

e. Area E:

   ____ unit × ____ units = ____ square units

f. Area F:

   ____ units × ____ units = ____ square units

2. The area of Benjamin's bedroom floor is shown on the grid to the right. Each ☐ represents 1 square foot. How many total square feet is Benjamin's floor?

   a. Label the side lengths.

   b. Use a straight edge to draw a grid of equal size squares within the rectangle.

   c. Find the total number of squares.

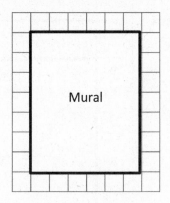

Benjamin's Bedroom Floor

3. Mrs. Young's art class needs to create a mural that covers exactly 35 square feet. Mrs. Young marks the area for the mural as shown on the grid. Each ☐ represents 1 square foot. Did she mark the area correctly? Explain your answer.

Mural

4. Mrs. Barnes draws a rectangular array. Mila skip-counts by fours and Jorge skip-counts by sixes to find the total number of square units in the array. When they give their answers, Mrs. Barnes says that they are both right.

   a. Use pictures, numbers, and words to explain how Mila and Jorge can both be right.

   b. How many square units might Mrs. Barnes' array have had?

Lesson 7:   Interpret area models to form rectangular arrays.

©2018 Great Minds®. eureka-math.org

Name _____   Date _____

1. Label the side lengths of Rectangle A on the grid below. Use a straight edge to draw a grid of equal size squares within Rectangle A. Find the total area of Rectangle A.

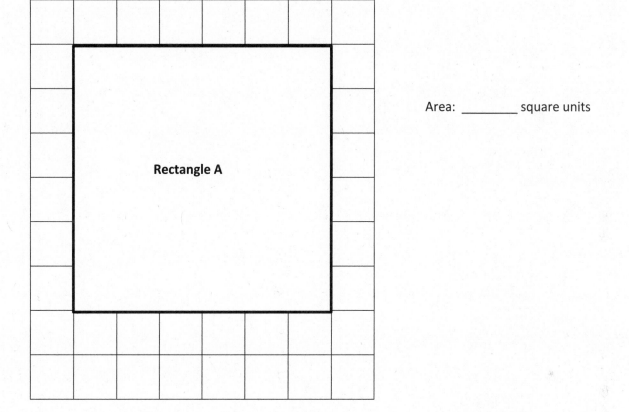

Area: _____ square units

2. Mark makes a rectangle with 36 square centimeter tiles. Gia makes a rectangle with 36 square inch tiles. Whose rectangle has a bigger area? Explain your answer.

area model

Marnie and Connor both skip-count square units to find the area of the same rectangle. Marnie counts, "3, 6, 9, 12, 15, 18, 21." Connor counts, "7, 14, 21." Draw what the rectangle might look like, and then label the side lengths and find the area.

Read   Draw   Write

Lesson 8:   Find the area of a rectangle through multiplication of the side lengths.

183

©2018 Great Minds®. eureka-math.org

Name _____   Date _____

1. Write a multiplication equation to find the area of each rectangle.

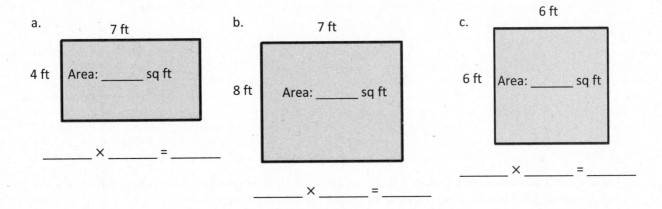

a.
7 ft
4 ft | Area: _____ sq ft

_____ × _____ = _____

b.
7 ft
8 ft | Area: _____ sq ft

_____ × _____ = _____

c.
6 ft
6 ft | Area: _____ sq ft

_____ × _____ = _____

2. Write a multiplication equation and a division equation to find the unknown side length for each rectangle.

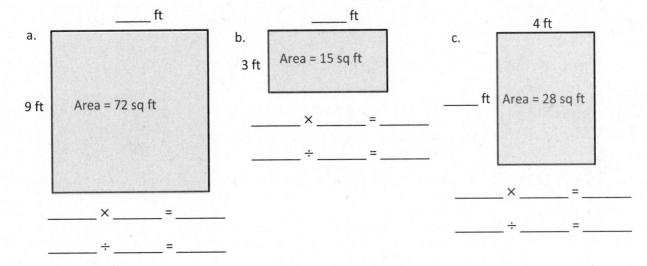

a.
_____ ft
9 ft | Area = 72 sq ft

_____ × _____ = _____

_____ ÷ _____ = _____

b.
_____ ft
3 ft | Area = 15 sq ft

_____ × _____ = _____

_____ ÷ _____ = _____

c.
4 ft
_____ ft | Area = 28 sq ft

_____ × _____ = _____

_____ ÷ _____ = _____

3. On the grid below, draw a rectangle that has an area of 42 square units. Label the side lengths.

4. Ursa draws a rectangle that has side lengths of 9 centimeters and 6 centimeters. What is the area of the rectangle? Explain how you found your answer.

5. Eliza's bedroom measures 6 feet by 7 feet. Her brother's bedroom measures 5 feet by 8 feet. Eliza says their rooms have the same exact floor area. Is she right? Why or why not?

6. Cliff draws a rectangle with a side length of 6 inches and an area of 24 square inches. What is the other side length? How do you know?

Lesson 8: Find the area of a rectangle through multiplication of the side lengths.

Name _____   Date _____

1. Write a multiplication equation to find the area of the rectangle below.

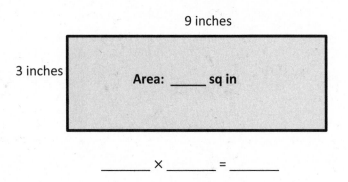

9 inches

3 inches   **Area:** _____ **sq in**

_____ × _____ = _____

2. Write a multiplication equation and a division equation to find the unknown side length for the rectangle below.

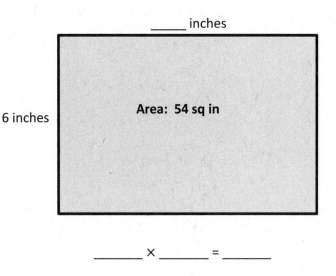

_____ inches

6 inches   **Area: 54 sq in**

_____ × _____ = _____

_____ ÷ _____ = _____

grid

EUREKA MATH™

Lesson 8:   Find the area of a rectangle through multiplication of the side lengths.

189

©2018 Great Minds®. eureka-math.org

Mario plans to completely cover his 8-inch by 6-inch piece of cardboard with square inch tiles. He has 42 square inch tiles. How many more square inch tiles does Mario need to cover the cardboard without any gaps or overlap? Explain your answer.

_____

_____

_____

Read   **Draw**   **Write**

Name _____   Date _____

1. Cut the grid into 2 equal rectangles.

    a. Draw and label the side lengths of the 2 rectangles.

    b. Write an equation to find the area of 1 of the rectangles.

    c. Write an equation to show the total area of the 2 rectangles.

2. Place your 2 equal rectangles side by side to create a new, longer rectangle.

    a. Draw an area model to show the new rectangle. Label the side lengths.

    b. Find the total area of the longer rectangle.

3. Furaha and Rahema use square tiles to make the rectangles shown below.

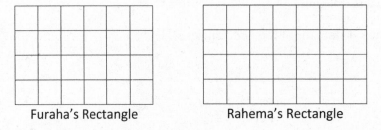

Furaha's Rectangle   Rahema's Rectangle

a. Label the side lengths on the rectangles above, and find the area of each rectangle.

b. Furaha pushes his rectangle next to Rahema's rectangle to form a new, longer rectangle. Draw an area model to show the new rectangle. Label the side lengths.

c. Rahema says the area of the new, longer rectangle is 52 square units. Is she right? Explain your answer.

4. Kiera says she can find the area of the long rectangle below by adding the areas of Rectangles A and B. Is she right? Why or why not?

Rectangle A   Rectangle B

        Lesson 9:   Analyze different rectangles and reason about their area.

©2018 Great Minds®. eureka-math.org

Name _____   Date _____

Lamar uses square tiles to make the 2 rectangles shown below.

Rectangle A   Rectangle B

1. Label the side lengths of the 2 rectangles.

2. Write equations to find the areas of the rectangles.

    Area of Rectangle A: _____   Area of Rectangle B: _____

3. Lamar pushes Rectangle A next to Rectangle B to make a bigger rectangle. What is the area of the bigger rectangle? How do you know?

Sonya folds a 6-inch by 6-inch piece of paper into 4 equal parts (shown below). What is the area of 1 of the parts?

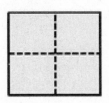

_____

_____

_____

_____

Read   **Draw**   **Write**

EUREKA MATH™

Lesson 10:   Apply the distributive property as a strategy to find the total area of a large rectangle by adding two products.

197

©2018 Great Minds®. eureka-math.org

Name _____ Date _____

1. Label the side lengths of the shaded and unshaded rectangles when needed. Then, find the total area of the large rectangle by adding the areas of the two smaller rectangles.

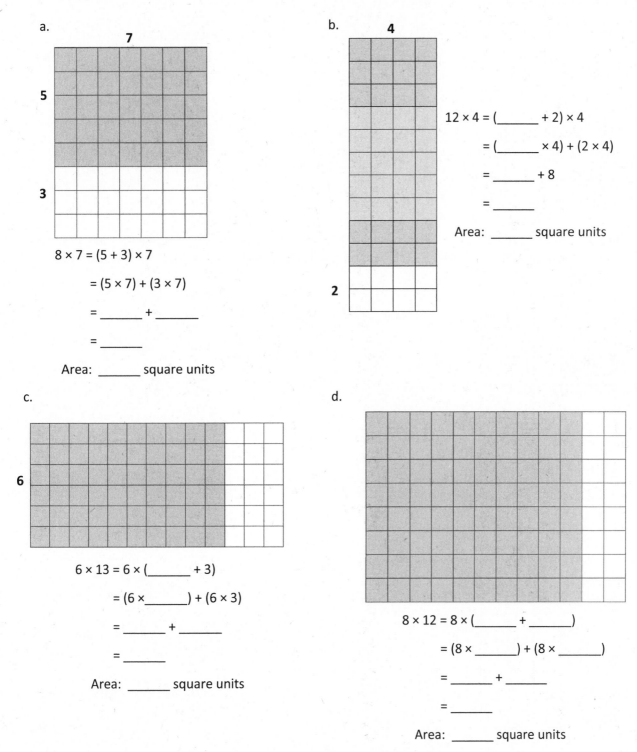

a.

7

5

3

$8 \times 7 = (5 + 3) \times 7$

$= (5 \times 7) + (3 \times 7)$

$= \underline{\hspace{1cm}} + \underline{\hspace{1cm}}$

$= \underline{\hspace{1cm}}$

Area: _____ square units

b.

4

2

$12 \times 4 = (\underline{\hspace{1cm}} + 2) \times 4$

$= (\underline{\hspace{1cm}} \times 4) + (2 \times 4)$

$= \underline{\hspace{1cm}} + 8$

$= \underline{\hspace{1cm}}$

Area: _____ square units

c.

6

$6 \times 13 = 6 \times (\underline{\hspace{1cm}} + 3)$

$= (6 \times \underline{\hspace{1cm}}) + (6 \times 3)$

$= \underline{\hspace{1cm}} + \underline{\hspace{1cm}}$

$= \underline{\hspace{1cm}}$

Area: _____ square units

d.

$8 \times 12 = 8 \times (\underline{\hspace{1cm}} + \underline{\hspace{1cm}})$

$= (8 \times \underline{\hspace{1cm}}) + (8 \times \underline{\hspace{1cm}})$

$= \underline{\hspace{1cm}} + \underline{\hspace{1cm}}$

$= \underline{\hspace{1cm}}$

Area: _____ square units

EUREKA MATH™

Lesson 10: Apply the distributive property as a strategy to find the total area of a large rectangle by adding two products.

199

©2018 Great Minds®. eureka-math.org

2. Vince imagines 1 more row of eight to find the total area of a 9 × 8 rectangle. Explain how this could help him solve 9 × 8.

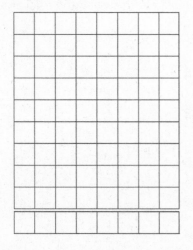

3. Break the 15 × 5 rectangle into 2 rectangles by shading one smaller rectangle within it. Then, find the sum of the areas of the 2 smaller rectangles and show how it relates to the total area. Explain your thinking.

Lesson 10:   Apply the distributive property as a strategy to find the total area of a large rectangle by adding two products.

©2018 Great Minds®. eureka-math.org

Name _____   Date _____

Label the side lengths of the shaded and unshaded rectangles. Then, find the total area of the large rectangle by adding the areas of the 2 smaller rectangles.

1.

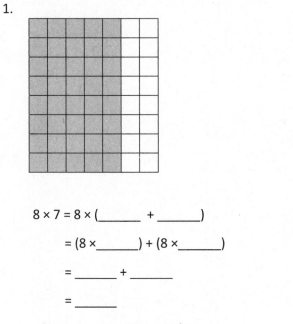

$8 \times 7 = 8 \times ($ _____ $+$ _____ $)$

$= (8 \times$ _____ $) + (8 \times$ _____ $)$

$=$ _____ $+$ _____

$=$ _____

Area: _____ square units

2.

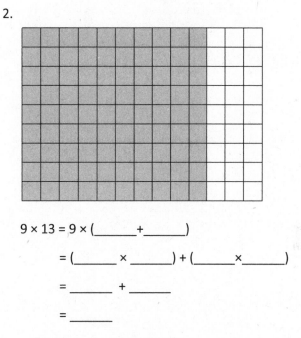

$9 \times 13 = 9 \times ($ _____ $+$ _____ $)$

$= ($ _____ $\times$ _____ $) + ($ _____ $\times$ _____ $)$

$=$ _____ $+$ _____

$=$ _____

Area: _____ square units

Lesson 10:   Apply the distributive property as a strategy to find the total area of a
             large rectangle by adding two products.

©2018 Great Minds®. eureka-math.org

201

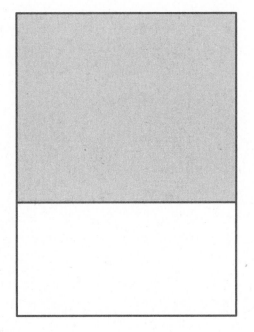

tiling

EUREKA MATH™

Lesson 10:   Apply the distributive property as a strategy to find the total area of a
large rectangle by adding two products.

203

©2018 Great Minds®. eureka-math.org

The banquet table in a restaurant measures 3 feet by 6 feet. For a large party, workers at the restaurant place 2 banquet tables side by side to create 1 long table. Find the area of the new, longer table.

Read   **Draw**   **Write**

EUREKA MATH™

Lesson 11:   Demonstrate the possible whole number side lengths of rectangles with areas of 24, 36, 48, or 72 square units using the associative property.

©2018 Great Minds®. eureka-math.org

205

Name _____   Date _____

1. The rectangles below have the same area. Move the parentheses to find the unknown side lengths. Then, solve.

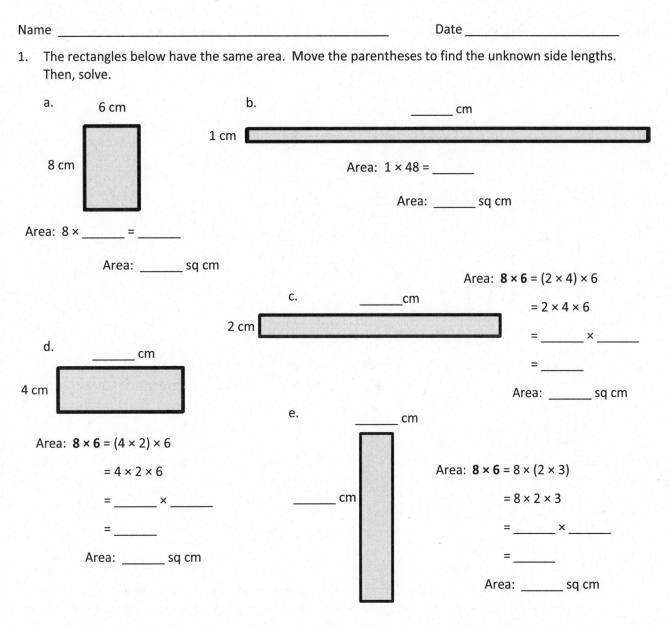

a.

6 cm

8 cm

Area: 8 × _____ = _____

Area: _____ sq cm

b.

_____ cm

1 cm

Area: 1 × 48 = _____

Area: _____ sq cm

c.

_____ cm

2 cm

Area: **8 × 6** = (2 × 4) × 6

= 2 × 4 × 6

= _____ × _____

= _____

Area: _____ sq cm

d.

_____ cm

4 cm

Area: **8 × 6** = (4 × 2) × 6

= 4 × 2 × 6

= _____ × _____

= _____

Area: _____ sq cm

e.

_____ cm

_____ cm

Area: **8 × 6** = 8 × (2 × 3)

= 8 × 2 × 3

= _____ × _____

= _____

Area: _____ sq cm

2. Does Problem 1 show all the possible whole number side lengths for a rectangle with an area of 48 square centimeters? How do you know?

Lesson 11:   Demonstrate the possible whole number side lengths of rectangles with areas of 24, 36, 48, or 72 square units using the associative property.

©2018 Great Minds®. eureka-math.org

207

3. In Problem 1, what happens to the shape of the rectangle as the difference between the side lengths gets smaller?

4. a. Find the area of the rectangle below.

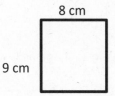

8 cm

9 cm

b. Julius says a 4 cm by 18 cm rectangle has the same area as the rectangle in Part (a). Place parentheses in the equation to find the related fact and solve. Is Julius correct? Why or why not?

$$4 \times 18 = 4 \times 2 \times 9$$

$$= 4 \times 2 \times 9$$

$$= \underline{\hspace{1cm}} \times \underline{\hspace{1cm}}$$

$$= \underline{\hspace{1cm}}$$

Area: _____ sq cm

c. Use the expression 8 × 9 to find different side lengths for a rectangle that has the same area as the rectangle in Part (a). Show your equations using parentheses. Then, estimate to draw the rectangle and label the side lengths.

Lesson 11:   Demonstrate the possible whole number side lengths of rectangles with areas of 24, 36, 48, or 72 square units using the associative property.
©2018 Great Minds®. eureka-math.org

EUREKA MATH

Name _____   Date _____

1. Find the area of the rectangle.

8 cm

8 cm

2. The rectangle below has the same area as the rectangle in Problem 1. Move the parentheses to find the unknown side lengths. Then, solve.

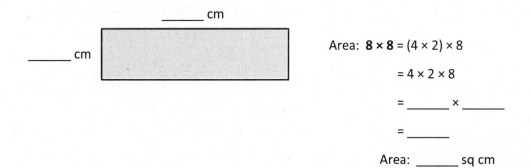

_____ cm

_____ cm

Area: **8 × 8** = (4 × 2) × 8

= 4 × 2 × 8

= _____ × _____

= _____

Area: _____ sq cm

a. Find the area of a 6 meter by 9 meter rectangle.

b. Use the side lengths, 6 m × 9 m, to find different side lengths for a rectangle that has the same area. Show your equations using parentheses. Then estimate to draw the rectangle and label the side lengths.

Read   Draw   Write

Name _____   Date _____

1. Each side on a sticky note measures 9 centimeters. What is the area of the sticky note?

2. Stacy tiles the rectangle below using her square pattern blocks.

   a. Find the area of Stacy's rectangle in square units. Then, draw and label a different rectangle with whole number side lengths that has the same area.

   b. Can you draw another rectangle with different whole number side lengths and have the same area? Explain how you know.

3. An artist paints a 4 foot × 16 foot mural on a wall. What is the total area of the mural? Use the break apart and distribute strategy.

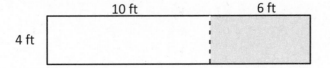

4. Alana tiles the 3 figures below. She says, "I'm making a pattern!"

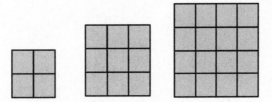

    a. Find the area of Alana's 3 figures and explain her pattern.

    b. Draw the next 2 figures in Alana's pattern and find their areas.

5. Jermaine glues 3 identical pieces of paper as shown below and makes a square. Find the unknown side length of 1 piece of paper. Then, find the total area of 2 pieces of paper.

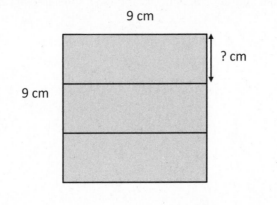

EUREKA MATH™

Name _____ Date _____

1. A painting has an area of 63 square inches. One side length is 9 inches. What is the other side length?

9 inches

Area = 63 square inches

2. Judy's mini dollhouse has one floor and measures 4 inches by 16 inches. What is the total area of the dollhouse floor?

Anil finds the area of a 5-inch by 17-inch rectangle by breaking it into 2 smaller rectangles. Show one way that he could have solved the problem. What is the area of the rectangle?

_____

_____

_____

_____

Read   **Draw**   **Write**

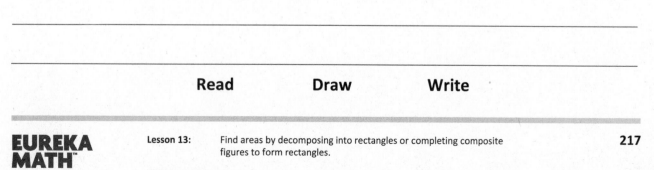

Lesson 13:   Find areas by decomposing into rectangles or completing composite figures to form rectangles.

217

©2018 Great Minds®. eureka-math.org

Name _____ Date _____

1. Each of the following figures is made up of 2 rectangles. Find the total area of each figure.

Figure 1: Area of A + Area of B: ____18____ sq units + _____ sq units = _____ sq units

Figure 2: Area of C + Area of D: _____ sq units + _____ sq units = _____ sq units

Figure 3: Area of E + Area of F: _____ sq units + _____ sq units = _____ sq units

Figure 4: Area of G + Area of H: _____ sq units + _____ sq units = _____ sq units

EUREKA MATH

Lesson 13: Find areas by decomposing into rectangles or completing composite figures to form rectangles.

219

©2018 Great Minds®. eureka-math.org

2. The figure shows a small rectangle cut out of a bigger rectangle. Find the area of the shaded figure.

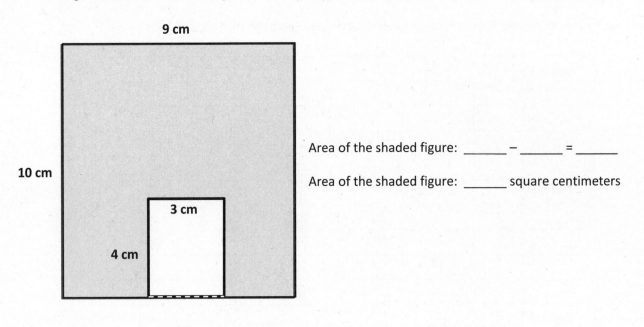

Area of the shaded figure: _____ – _____ = _____

Area of the shaded figure: _____ square centimeters

3. The figure shows a small rectangle cut out of a big rectangle.

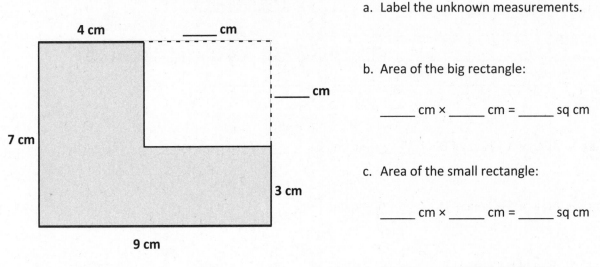

a. Label the unknown measurements.

b. Area of the big rectangle:

_____ cm × _____ cm = _____ sq cm

c. Area of the small rectangle:

_____ cm × _____ cm = _____ sq cm

d. Find the area of the shaded figure.

Lesson 13: Find areas by decomposing into rectangles or completing composite figures to form rectangles.

EUREKA MATH

Name _____ Date _____

The following figure is made up of 2 rectangles. Find the total area of the figure.

A

B

Area of A + Area of B: _____ sq units + _____ sq units = _____ sq units

Lesson 13: Find areas by decomposing into rectangles or completing composite figures to form rectangles.

221

©2018 Great Minds®. eureka-math.org

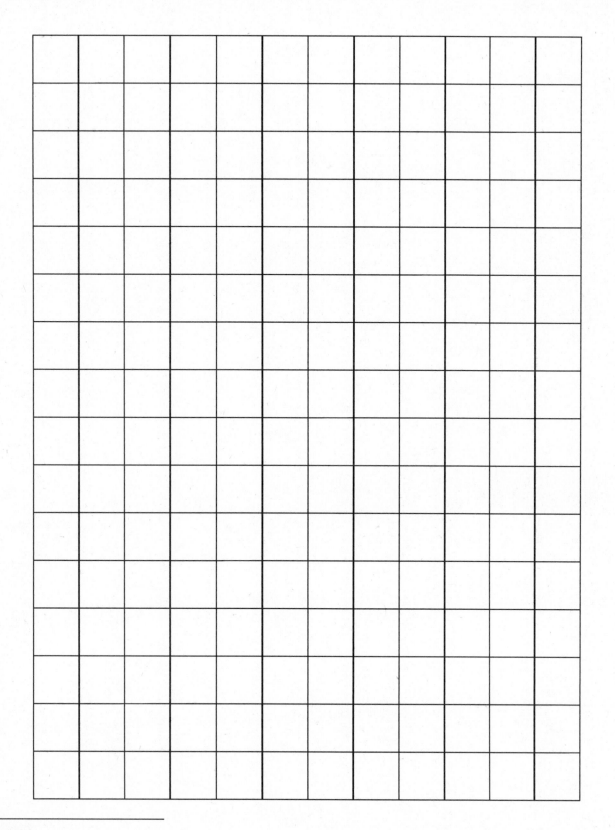

large grid

Lesson 13: Find areas by decomposing into rectangles or completing composite figures to form rectangles.

223

©2018 Great Minds®. eureka-math.org

a. Break apart the shaded figure into 2 rectangles. Then, add to find the area of the shaded figure below.

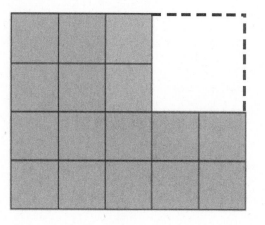

b. Subtract the area of the unshaded rectangle from the area of the large rectangle to check your answer in Part (a).

Read   **Draw**   **Write**

Lesson 14:   Find areas by decomposing into rectangles or completing composite
              figures to form rectangles.

225

©2018 Great Minds®. eureka-math.org

Name _____   Date _____

1.   Find the area of each of the following figures. All figures are made up of rectangles.

a.

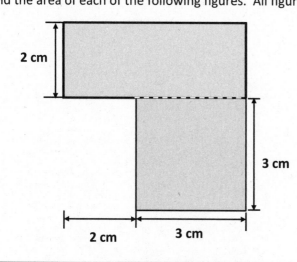

b.

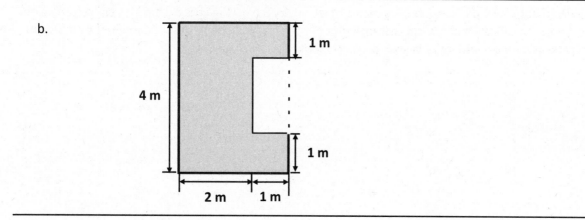

2.   The figure below shows a small rectangle in a big rectangle. Find the area of the shaded part of the figure.

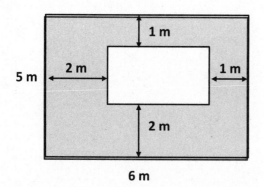

Lesson 14:   Find areas by decomposing into rectangles or completing composite figures to form rectangles.

©2018 Great Minds®. eureka-math.org

227

3. A paper rectangle has a length of 6 inches and a width of 8 inches. A square with a side length of 3 inches was cut out of it. What is the area of the remaining paper?

4. Tila and Evan both have paper rectangles measuring 6 cm by 9 cm. Tila cuts a 3 cm by 4 cm rectangle out of hers, and Evan cuts a 2 cm by 6 cm rectangle out of his. Tila says she has more paper left over. Evan says they have the same amount. Who is correct? Show your work below.

EUREKA MATH

Name _____   Date _____

Mary draws an 8 cm by 6 cm rectangle on her grid paper. She shades a square with a side length of 4 cm inside her rectangle. What area of the rectangle is left unshaded?

Lesson 14:   Find areas by decomposing into rectangles or completing composite
                  figures to form rectangles.

©2018 Great Minds®. eureka-math.org

229

Name _____ Date _____

1. Make a prediction: Which room looks like it has the biggest area?

2. Record the areas and show the strategy you used to find each area.

Room	Area	Strategy
Bedroom 1	_____ sq cm	
Bedroom 2	_____ sq cm	
Kitchen	_____ sq cm	
Hallway	_____ sq cm	
Bathroom	_____ sq cm	
Dining Room	_____ sq cm	
Living Room	_____ sq cm	

Lesson 15: Apply knowledge of area to determine areas of rooms in a given floor plan.

©2018 Great Minds®. eureka-math.org

231

3. Which room has the biggest area? Was your prediction right? Why or why not?

4. Find the side lengths of the house without using your ruler to measure them, and explain the process you used.

    Side lengths: _____ centimeters and _____ centimeters

5. What is the area of the whole floor plan? How do you know?

    Area = _____ square centimeters

The rooms in the floor plan below are rectangles or made up of rectangles.

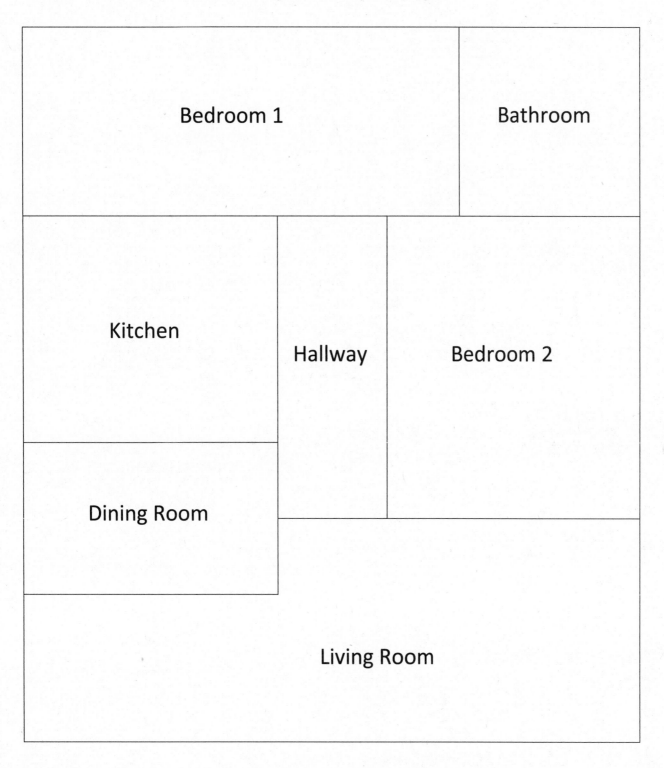

Lesson 15:   Apply knowledge of area to determine areas of rooms in a given floor
plan.

©2018 Great Minds®. eureka-math.org

233

Name _____ Date _____

Jack uses grid paper to create a floor plan of his room. Label the unknown measurements, and find the area of the items listed below.

Name	Equations	Total Area
a. Jack's Room		_____ square units
b. Bed		_____ square units
c. Table		_____ square units
d. Dresser		_____ square units
e. Desk		_____ square units

EUREKA MATH™

Lesson 15: Apply knowledge of area to determine areas of rooms in a given floor plan.

©2018 Great Minds®. eureka-math.org

235

Name _____ Date _____

Record the new side lengths you have chosen for each of the rooms and show that these side lengths equal the required area. For non-rectangular rooms, record the side lengths and areas of the small rectangles. Then, show how the areas of the small rectangles equal the required area.

Room	New Side Lengths
Bedroom 1: 60 sq cm	
Bedroom 2: 56 sq cm	
Kitchen: 42 sq cm	

Lesson 16: Apply knowledge of area to determine areas of rooms in a given floor plan.

©2018 Great Minds®. eureka-math.org

237

Room	New Side Lengths
Hallway: 24 sq cm	
Bathroom: 25 sq cm	
Dining Room: 28 sq cm	
Living Room: 88 sq cm	

Lesson 16: Apply knowledge of area to determine areas of rooms in a given floor plan.

©2018 Great Minds®. eureka-math.org

Name _____   Date _____

Find the area of the shaded figure. Then, draw and label a rectangle with the same area.

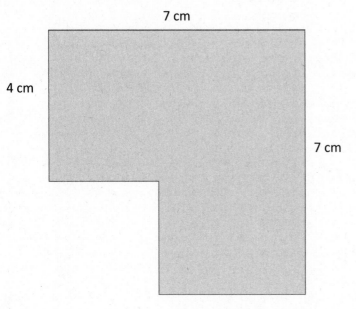

7 cm

4 cm

7 cm

4 cm

Lesson 16:   Apply knowledge of area to determine areas of rooms in a given floor
              plan.

239

©2018 Great Minds®. eureka-math.org

Credits

Great Minds® has made every effort to obtain permission for the reprinting of all copyrighted material. If any owner of copyrighted material is not acknowledged herein, please contact Great Minds for proper acknowledgment in all future editions and reprints of this module.